MAKING INNOVATION HAPPEN

Concept Management Through Integration

MAKING INNOVATION HAPPEN

Concept Management Through Integration

Gerhard Plenert
Shozo Hibino

St. Lucie Press
Boca Raton   Boston   London   New York   Washington, D.C.

Library of Congress Cataloging-in-Publication Data

Plenert, Gerhard Johannes.
    Making innovation happen : concept management through integration / Gerhard Plenert.
Shozo Hibino : illustrations by Gerick Plenert.
        p.   cm.
    Includes bibliographical references and index.
    ISBN 1-57444-090-X (alk. paper)
    1. Total quality management.   2. Teams in the workplace.
  3. Leadership.   4. Production management.   I. Hibino, Shozo, 1940–
    II. Title.
  HD62.15.P56   1997
  658.5--dc21   97-37096
                                                                                                          CIP

This book contains information obtained from authentic and highly regarded sources. Reprinted material is quoted with permission, and sources are indicated. A wide variety of references are listed. Reasonable efforts have been made to publish reliable data and information, but the author and the publisher cannot assume responsibility for the validity of all materials or for the consequences of their use.

Neither this book nor any part may be reproduced or transmitted in any form or by any means, electronic or mechanical, including photocopying, microfilming, and recording, or by any information storage or retrieval system, without prior permission in writing from the publisher.

The consent of CRC Press LLC does not extend to copying for general distribution, for promotion, for creating new works, or for resale. Specific permission must be obtained in writing from CRC Press LLC for such copying.

Direct all inquiries to CRC Press LLC, 2000 Corporate Blvd., N.W., Boca Raton, Florida 33431.

© 1998 by CRC Press LLC
St. Lucie Press is an imprint of CRC Press LLC

No claim to original U.S. Government works
International Standard Book Number 1-57444-090-X
Library of Congress Card Number 97-37096
Printed in the United States of America 1 2 3 4 5 6 7 8 9 0
Printed on acid-free paper

Dedication

To my family, especially my wife, Renee Sangray Plenert, whose challenges force continuous innovation into my life. And to my God, who tells me through the scriptures that I cannot get better without dramatic changes.

Gerhard Johannes Plenert

To my long-term adviser, Dr. Gerald Nadler, with whom I co-created a New Thinking Paradigm, and to my wife, Shigemi Hibino, who supported my research and consulting work throughout my life. My fruitful life would not exist without them.

Shozo Hibino

Table of Contents

About the Authors ix
Foreword xi
Preface xiii
Acknowledgments xv
Introduction xvii

PART I: TRADITION

1. Why Change? 3
2. Creative (Breakthrough) Thinking 9
3. World Class Management 29
4. Change Methodologies 51

PART II: CONCEPT MANAGEMENT

5. Concept Management Philosophy 81
6. Concept Management — Integration and Organization: "How CM Works" 99
7. Concept Management: Measurement and Motivation 119
8. Concept Management: Working Together 129
9. Concept Management: The Road to Success 143

PART III: A NEW WORLD ORDER OF CHANGE

10. Concept Management: The Test 153
11. Concept Management In Your Future 159

Index 165

About the Authors

Gerhard Plenert, Ph.D.
President, Institute of World Class Management
Director of Quality
Precision Printers, Inc.
6624 Penney Way
Carmichael, CA 95608

Phone (916) 536-9751
Fax (916) 536-9758
E-mail: plenert@aol.com

Gerhard Plenert, Ph.D., CPIM, spent 20 years working in management in private industry. His specialty was MIS (Management Information Systems) and POM (Production/Operations Management). He traveled throughout the world in this capacity teaching industrial management and solving industrial problems. He lived in Mexico 1½ years. He returned to Colorado School of Mines to earn his Ph.D. He then spent 5 years teaching and researching at California State University, Chico. He joined Brigham Young University in the Institute of Business Management, where he has been teaching and researching for the last 6 years. He was the Director of the *California Productivity and Quality Center* at CSUC and is now the Director of the *Productivity and Quality Research Group* at BYU. His research specialty is in International Industrial Management, with emphasis on developing countries. He spent 1 year in Malaysia and has made numerous trips to other developing countries. He has published three books: *International Management and Production — Survival Techniques for Corporate America*, an APICS series book titled *Plant Operations*

Handbook and, most recently, *World Class Manager*. He has published over 110 articles and has presented over 140 talks/seminars during his academic career.

Shozo Hibino, Ph.D.
President, The Japan Planology Society
Chukyo University
School of Sociology
101 Tokodachi, Kaizu-Cho
Toyota, Japan 470-03

Phone 81-52-793-1417
Fax 81-52-796-3956
E-Mail: shibino@cnc.chukyo-u.ak.jp

Shozo Hibino, Ph.D., is Professor of Planning and Design at Chukyo University, Japan, is President of the Japan Planology Society, and has over 25 years of academic and consulting experience. He has worked for small and medium-sized businesses, including the Junior Chamber of Commerce, and also for large industries, such as Mitsubishi, Canon, NEC groups, Fuji Xerox, Chubu Electric Power, Hotel Nagoya Castle groups, Toyota groups, Tokai Bank, and a variety of other types of industries in a variety of different countries.

He has published over 100 articles and 20 books. His most recent books include *Breakthrough Thinking, Breakthrough, Breakthrough Re-Engineering, Breakthrough QC,* and *Creative Solution Finding*. The first, *Breakthrough Thinking* with Gerald Nadler, was published in many different languages and has become a best-seller worldwide.

He received the Phi Kappa Phi Faculty Recognition Award in 1990 and the Ban Memorial Award in 1993.

Foreword

Management publications at last are describing the continual and heavy flow of "savior" program solutions in the way they should have been labeled from the beginning — fads, quick fixes, alphabets *du jour*, and vapid superficialities. John Micklethwait and Adrian Wooldridge, in the book *The Witch Doctors,* note that each one in the parade of *the answers* is given "some acronym and tarted up in scientific language" and "profitably" stimulates the management-advice sector. They list, among many others, "re-engineering, ... management by objectives (MBO), ... T-groups, ... management by walking around, [and] intrapreneurship."

Scott Adams is extremely successful in poking fun, through his *Dilbert* comics, at the top executives who so willingly adopt these fads without serious consideration of the consequences of each program and of the foolish pictures of themselves the succession of "answers" provides.

Each program in almost every case does have one or two worthwhile goals (out of the whole set an organization needs to consider) — quality, measurable objectives (e.g., cost reduction), employee involvement, management style, lead time, and creativity, among them. But more seriously, in my mind, each one is accompanied by a whole list of "thou-shalts and shalt nots" and a "new" type of organizational structure that is "necessary" to achieve its goals. This churning in the way the company operates according to the dictates of an external program is the cause of the jaded, morale-sapping, and lack-of-respect attitude of almost everyone in the business world (except, of course, the perpetrators of the twists and turns).

The last 10 to 12 years have seen the "positive" results of these fads — making money by cutting costs (with its attendant human travails). But this result has led to the many problems recognized by serious reviewers of the scene, among the most important being how to develop new products, processes, and services that are so critical to the growth and long-term success of the organization. (Organizational changes to accommodate such needs, however frequent, are much different than the churning due to "programs.") Interestingly, recognizing the need for such creativity and innovation has spawned a whole new set of faddish programs — artist as creativity model, living organisms as surrogates for the business, learning organization, and decentralization, to name a few. Some good goals, yes, but do they change anything in the mix of difficulties I identified before?

Plenert and Hibino move in the right direction in *Making Innovation Happen*. They seek to integrate a different approach to creating and restructuring systems and solutions for producing significant organizational results while incorporating the values and goals of several programs. They recognize the need to use different principles as the basis for how to think. They show how the new thinking action steps are different. They propose that these differences could produce effective results. The subtitle of the book, "Concept Management Through Integration," displays the intent to provide some grounded perspectives that go beyond the acronym and fad.

Read on. Your introduction to the ideas here should stimulate you to revise your way of going about getting results from your change efforts.

Gerald Nadler
March 1997
Los Angeles, CA

Preface

I am always doing what I can't do yet in order to learn how to do it.

Vincent Van Gogh

The goal of every successful management program should be to help readers become better at satisfying their customers. This is accomplished by maximizing human potential through organized, integrated systems that focus on developing creativity and innovation. The implementation and integration of these success systems is known as Concept Management, hence the title of this book. It is not that the title contains any new words, but the words in the title generate a whole new world of ideas. Competitiveness in today's world demands innovative ideas. These ideas have no value if they are not implemented as changes throughout the organization. These concepts are a blend of current world thought, pulling the best from many worlds to create an environment of exciting, purpose-goal-directed, positive, innovative change.

The focus (mission) of this book is fourfold:

1. Integration (World Class Management)
2. Innovation (Concept Management)
3. Creativity — idea generation (Breakthrough Thinking®)
4. Quality and productivity improvements in the area of manufacturing management output, as demonstrated through stories and examples

The "integration" occurs because several leading-edge concepts are being brought together into one package:

1. Breakthrough Thinking (Nadler and Hibino)
2. World Class Management (Plenert)
3. Total Quality Management with Concept Engineering

This integration brings together the leading-edge management philosophies from both sides of the Pacific: Japan in the East and the United States in the West. It also integrates areas from North and South, including the innovative ideas of manufacturers from Brazil, Malaysia, and Australia.

This book offers

1. A new, advanced change philosophy
2. An integration of change methodologies
3. An integration of Breakthrough Thinking and World Class Management
4. References for further study of the concepts introduced
5. Foreword by Gerald Nadler (coauthor in the Breakthrough Thinking series)
6. A quality and manufacturing management focus

The last thing we need is another "management book," filled with traditional philosophies about how to manage an organization. This book provides the reader with nontraditional, innovative, breakthrough thinking, and world class ideas that are leading edge. That is exactly what *Making Innovation Happen: Concept Management Through Integration* is designed to do. It takes the leading-edge philosophies and management tools from both the East and the West and generates a synergistic, integrated approach to change that will not just conquer, but stun the competition. It is impossible to go through the ideas and concepts in this book and not come away with processes and methodologies that you haven't used before. The integration of these different concepts is a new and enormously competitive approach to doing business that reaches well out into the twenty-first century. It is a book designed to show how to break from traditional methodologies and thinking and find new, competitive avenues of success within your own organization.

Acknowledgments

Credit for this book is due to a long and unending list of people, starting with our families, who supported us through this adventure in innovation. We thank California State University, Chico, and Chukyo University, which introduced us and offered us our first chance to work together on ideas and articles. Thanks also goes to Brigham Young University, which supported our joint research efforts on this book. And thanks to Dr. Gerald Nadler who gave us helpful advice in making this piece of innovation happen.

Thanks is given to the many consulting companies that provided some of the cases in Concept Management. Last of all, we give thanks to Drew Gierman, of St. Lucie Press, who patiently worked with us until we finally came up with a product that was both usable and marketable.

Shozo Hibino
Gerhard Plenert

Introduction

The definition of insanity is continuing to do the same things and expecting different results.

Breakthrough Thinking, Nadler and Hibino

If you put a frog in a frying pan and turn up the heat, the frog will jump without first carefully considering to where it is jumping. In this fable, the heat is change, and we are the frogs. When the heat gets turned up, we jump, but sometimes we jump and land right back in the pan. Sometimes we jump and end up in the fire. Occasionally, if we are lucky, we may jump safely; but wouldn't it be smarter to be controlling the temperature, thereby allowing us to decide when and where to jump? Then we have a better chance of ending up in safety. That's the difference between managing change and letting change manage us. If we manage change, then we can decide

> When to jump
> Where to jump
> How to jump
> What is our safety zone for jumping
> What will be our agent for jumping

However, if we are the frying-pan frog, we lose control of our ability to make jump decisions, and we let the pressure of the moment make the jump decision for us. Or worse yet, maybe we're stubborn enough to sit and get fried when the heat gets turned up.

This book is the culmination of two leading-edge philosophies focused on Change Management. The first is "Breakthrough Thinking" (from the book of the same name by Nadler and Hibino), which focuses away from the traditional backward-looking approach to change toward innovative, creative, forward-looking change. This new book focuses on making innovation happen by utilizing "Breakthrough Thinking" (BT), an opportunity-generating paradigm.

The second leading-edge philosophy in this book is "World Class Management" (WCM) (from the book of the same name by Gerhard Plenert), which stresses that managers need to make change happen, rather than make changes only because they are forced to. Managers need to be the source of innovation. Both philosophies, BT and WCM, stress the need for change. BT offers the "how to" of identifying innovative change; WCM offers the "where to" change and shows us how to proceed with the change (innovation) implementation process.

Making Innovation Happen: Concept Management Through Integration formalizes the change (innovation) process. It recognizes that most individuals don't make change happen because

1. they are afraid of how change will affect them, or
2. they don't know how to make change happen.

For example, in a large corporate setting, more time is spent arguing about what to change and how to initiate the change than on the actual implementation. This new book puts structure into the change process so that everyone, at all levels of the organization, understand

Introduction

What needs changing
How to identify innovative alternatives for change
How to select the best change alternative
How to initiate the change process
How to monitor and control the change process
How to measure the performance of the change
How to establish an ongoing feedback mechanism under which the change operates

This book looks at the existing change models, such as Total Quality Management (TQM), Process Reengineering (PR), ISO 9000, and the award processes such as the Deming Award, the Baldrige Award, and the Shingo Prize. It also looks at the latest quality programs of companies such as the Japanese divisions of Toyota, Xerox, Canon, and Mitsubishi, as well as American companies of AT&T, Motorola, and FedEx. The result is a new change (innovation) management tool called Concept Management.

The title of this book, *Making Innovation Happen: Concept Management Through Integration,* was chosen from a dozen different options. There is significance in this title, and we would like to share why. "Innovation" expresses clearly that we are looking for change, but not any kind of change; we are looking specifically for creative, positive, non-copy-cat, purpose-goal-directed change that moves us out and beyond the competition. "Making Innovation Happen" expresses our desire to develop innovative change mechanisms in the individuals that read this book. "Concept Engineering" is a buzzword used in Japan to identify the innovative, creative process, but we were looking for more than just innovation, we were looking for the managed, purpose-goal-directed development of innovation, which is expressed by the term "Concept Management." Last, "Integration" expresses the need for an organizational initiative toward the change process. The effectiveness of a change program is greatly enhanced if there is an integrated push toward results.

This book provides a tool for formalizing innovation in general management; its focus is on manufacturing and quality management. The book will help you identify the need and methods for change, and will also provide a medium for the innovative change process. It will stimulate change by helping to understand and simplify the change process. It will resolve resistance and fear by providing a conceptual tool for identifying innovation and implementing change.

TRADITION 1

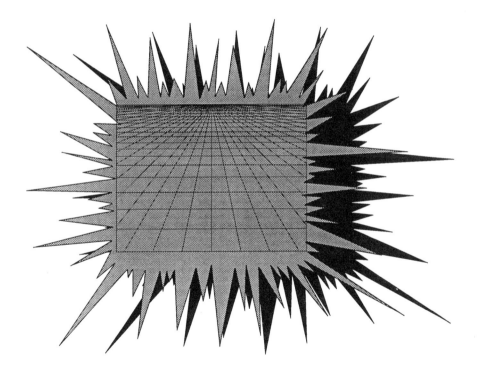

Chapter 1

Why Change?

> *What can be worse than being blind?*
> *Having eyes to see, but no vision.*
>
> **Helen Keller**

If you want to cook a frog, you can

1. Put him in hot water — he'll jump right back out
2. Put him in cold water and quickly heat the water — he'll notice the change and jump out
3. Put him in cold water and slowly heat the water — he won't even realize what's happening to him until it's too late and he's cooked!

Most change, especially competitive change, happens slowly and subtly. Before we realize what's happening, we're cooked. We can be cooked by losing our competitive edge, or by not being ready for technological advances. We need to be ready for change by preparing for it, by being ahead of it, not by letting it be forced upon us. The company that plans today for its competitive position 5 years from now will be ahead of the pack when the 5 years are up.

There are many reasons (causes, excuses, sources) for change, such as external (uncontrolled) and internal (controllable) change. We have no control over some changes, such as earthquakes; however, we have control over how prepared we are for them and how we react. Concept management helps us prepare for and manage the unexpected.

Other changes, such as technological advances or competitive position, are things we have direct control over, and it is up to us to successfully

act as change agents, watching for and creating opportunities. We need to be able to coordinate these changes through Concept Management.

Many factors trigger change, such as human awareness, human values, and human motivation; and all of these factors can be refocused toward a new perception of change, a new viewpoint. However, we need to know what to look for. We need a set of glasses that give us the "big picture," for us to see clearly through Concept Management tools.

Doyle Wilson, president and CEO of Doyle Wilson Homebuilder, Inc., of Austin, Texas, has demonstrated an impressive ability to implement Concept Management. His vision is focused toward positive, growth-oriented change. He believes that with forward vision you can improve quality, productivity, and efficiency in any company, even when tradition says that you are in a "slow industry."[1] Doyle Wilson's "continuous improvement" strategy introduced quality and time-to-market innovation into the home construction industry. He has reduced the construction cycle time for a new home from 165 days to 124 days and is still cutting off more. The cost reduction of this 25% faster cycle time allows him to sell his homes for as much as 10% less than the competition. Additionally, he uses Deming quality philosophies and Toyota lean manufacturing principles, applying them to his construction process. These principles have further assisted in the reduction of cycle-time and costs. Wilson's creativity has won him the coveted National Housing Quality Award. It is Wilson's style of creativity and innovation (Concept Management) that is the focus of this book. This book hopes to make us all "Doyle Wilsons" of our own industries.

Why Change?

The purpose of this book is to

1. Show us how to make leaps out of the frying pan and still land safely
2. Help us recognize when the water is getting warmer so we know what to change and when to change it
3. Help us optimize human potential through planned change processes
4. Help us create satisfied customers
5. Help us manage the change process (Concept Management) by:
   Identifying what our target objectives (ideal solutions) are, and
   Developing through innovation and creativity a road map showing where we are and laying out a plan for getting from this starting point toward finally achieving our target objective.

Traditional models of change, such as the scientific method or TRIZ,[2] focus on "push" thinking, where we analytically attack a problem, often incorporating extensive amounts of data collection and information analysis. In push thinking we are looking for the problem's "root cause" and we are pushing a solution at this cause. Conversely, with Concept Management, we are focusing on the ideal target solution (objective) and we allow this objective to "pull" us toward our objective. Where we are now is not as critical as where we want to end up. Therefore, in the management of the change process, identifying where we are now is of minor importance, as opposed to the pull system, where identifying where we are now is the subject of intensive data analysis.

In order to develop this process of creativity, we have found it necessary to create a merger of two leading-edge philosophies: World Class Management[3] and Breakthrough Thinking.[4] The merger works as follows:

Defining Target Objectives (ideal solutions):
   Breakthrough Thinking, through Purpose Expansion, helps develop creative, innovative goals (target solutions).
   World Class Management helps formalize those goals into a structure that outlines the focus and format of the target.
Drawing the Road Map:
   Breakthrough Thinking offers a new vision of the road map's purpose, and offers a methodology that will make the trip shorter, with fewer bumps.
   World Class Management offers the structure to design the road map. It locates where we are now and helps us build a map toward our target solution.

Ricardo Semler, CEO of Semco in Brazil,[5] identified what he wanted his organization to become and developed a road map, through his employees, of how he was going to get there. He wasn't happy with the tension and conflict in his company, and he wanted to change. His drive for change created a manufacturing company that has no receptionists, secretaries, traditional organizational charts, executive perks, or dress codes. Employees set their own work hours, define their own salaries, have access to all financial statements, set their own production quotas, make corporate decisions, redesign how products look and how they are built, and evaluate the performance of their managers. Employees are allowed, and often encouraged, to work at home. They are encouraged to set up their own businesses using company assets. Traditional computerization and cost-oriented financial analysis has been eliminated.

Semler says that the standard policy for the company is to have "no policy." In spite of this organizational plan, which defies all traditional logic, Semco has been enormously successful. Over a 10-year period during which the Brazilian economy was disastrous for business, Semco achieved a 600% growth rate. Productivity has increased nearly sevenfold. Profits have increased fivefold. Semco has gone as much as 14 months during which not one employee left the company.

What created this innovation for change at Semco? It's not that Ricardo Semler didn't understand how to run a business; initially he installed computers and developed financial and cost-analysis systems. He incorporated security systems, and used a very authoritarian management style, but he decided that his company just didn't "feel happy." He threw out all traditional methodologies. He felt that his employees knew more about his business than he did, and so they should be running it, and he turned it over to them. The change and innovation process started slowly, as change often does, and employees were sceptical. Today, Semco is an internationally recognized model for the successful implementation of teaming, empowerment, gainsharing, and world-class innovativeness. That is what Concept Management is all about.

Concept Management uses the term "concept" to mean innovative, purpose-driven change creation, and "management" to mean leadership. Therefore, concept management is "innovative, change-oriented, purpose-driven (goal focused), creative leadership." This leadership occurs through the integration of ideas, primarily the ideas expressed in two leading-edge philosophies: Breakthrough Thinking and World Class Management.

Realizing that the focus of all competitive issues, such as supply chain management or time-to-market effectiveness, is through leading-edge, positive change, we next need to see how change can best be identified (concept creation) and managed (concept management). In order to have

Why Change?

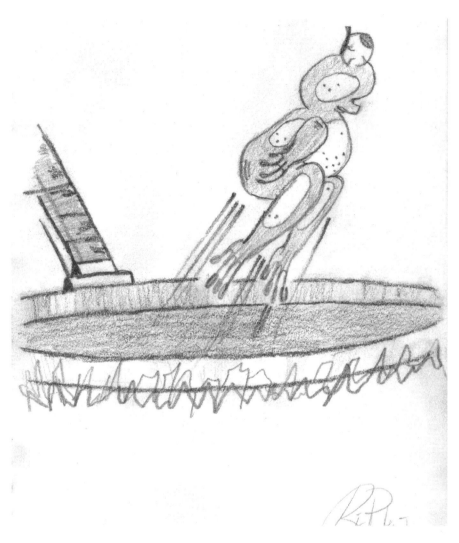

successful concept management, which is the road map for successful change implementation, we first need to go through a phase of concept creation, or change identification (defining our target solutions). Concept creation should be focused on value-added thinking. This book will give you the tools necessary for concept creation, and then assist you through the process of concept management. It will help you to use "pull" thinking, focusing on the future, rather than "push" thinking, focused on managing the past. It will teach you how to manage change by generating and managing future focused concepts.

This book starts with a discussion of Breakthrough Thinking and World Class Management, then discusses current change implementation methodologies such as Total Quality Management and Process Reengineering. These tools are integrated through a series of six chapters, which demonstrate how these techniques can be utilized. The book then summarizes the Concept Management process by showing how to initiate implementation. So let's get started toward leading-edge competitiveness. Let's jump out of the frying pan before we get cooked!

We miss 100% of the shots we don't take!

Endnotes

1. Novicki, Christina, "Meet the Best Little House Builder in Texas," *Fast Company*, August/September 1996, p. 38, 40. Additional information on Doyle Wilson's innovative processes can be found at his Web Site: http://www.doylewilson.com
2. Fey, Victor and Eugene Rivin, "TRIZ: A New Approach to Innovative Engineering and Problem Solving," *Target,* September/October 1996, p. 7–13.
3. The primary source for World Class tools and techniques is in Plenert, Gerhard, *World Class Manager,* Prima Publishing, Rocklin, CA, 1995.
4. The primary sources for Breakthrough Thinking tools and techniques are Nadler, Gerald and Shozo Hibino, *Breakthrough Thinking*, Prima Publishing, Rocklin, CA, 1994, Nadler, Gerald, and Shozo Hibino with John Farrell, *Creative Solution Finding*, Prima Publishing, Rocklin, CA, 1995, and Hoffherr, Glen D., John W. Moran, and Gerald Nadler, *Breakthrough Thinking in Total Quality Management*, PTR Prentice Hall, Englewood Cliffs, NJ, 1994.
5. Ricardo Semler wrote his own book on his experiences at Semco. The book is Semler, Ricardo, *Maverick*, Warner Books, New York, 1993.

Chapter 2

Creative (Breakthrough) Thinking

Fix the Problem, Not the Blame.

Anonymous

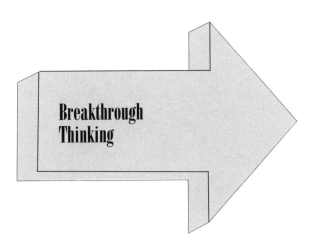

A frog and a monkey are traveling through the jungle. The frog hops along the ground and when he encounters a tree he looks for a way to get around that tree. However, the monkey, encountering the same tree, will not look at the tree as an obstacle to get around. Rather, the monkey will climb up into the tree and will look at the broader perspective. He will look at where he wants to go, and what is the best route to get there.

Using the frog's method, the frog may eventually encounter a cliff that will stop forward progress. However, the monkey will have noticed the cliff from the top of the tree and will have managed his way around it long before the cliff becomes an obstacle to forward progress.

The monkey sees the big picture, whereas the frog sees the inconvenient obstacles. Breakthrough Thinking gives us the ability to have a monkey's perspective. It lets us look down from the tree and envision the greater objective. Whereas frog management keeps us fighting the small fires, one at a time. Breakthrough Thinking gets us off of the ground and into the trees, offering us a full spectrum of creative solution finding.

> *We cannot talk people into accepting the future*
> *if they haven't been there in thought.*
> **Breakthrough Thinking Workbook**

Like the frog, our society, industry, and families are facing big obstacles. We resist jumping out into this world of problems. We have challenges, such as endless wars, environmental threats, a rapid increase of drugs, crime, gangs, the collapse of the family, traffic congestion, urban overcrowding, and so on.

In today's world we need a full spectrum of creative solution finding. Probably the most famous Nobel prize winner, Albert Einstein, said "we thus drift toward unparalleled catastrophe. We shall require an entirely new way of thinking, if we are to survive. ... Problems cannot be solved by applying the same type of thinking that created them."[1] To solve difficult problems and find creative solutions, our present thinking paradigm and process must change.

Gerald Nadler and Shozo Hibino published *Breakthrough Thinking*[2] in 1990 and *Creative Solution Finding*[3] in 1993. In these two books, they proposed a "Paradigm Shift in Thinking."[4] They called this new thinking paradigm "Breakthrough Thinking." The theory behind the paradigm is detailed in Parts I and II of *Creative Solution Finding*. Thinking paradigms have shifted over our history, for example, Primitive, Early Greek, Classical Greek, God Thinking, Descartes, etc. From the historical viewpoints, our thinking paradigms have been continuously shifting over time. Our conventional thinking paradigm (Descartes thinking) is out of date with a rapidly changing world and needs to shift again to a new thinking paradigm, Breakthrough Thinking. In the twenty-first century, we have to be Multi-Thinkers who are able to use three thinking paradigms; God Thinking, Conventional (Descartes) Thinking, and Breakthrough Thinking.

God Thinking starts with God's will. When we make a decision, we decide our behavior based on God's will. For some decisions, there is no need for analysis. Behavior is firmly dictated by God's will. For example, for the true believers, morality or ethical issues are decided and are not open for discussion. Today, many people utilize God Thinking.

Conventional Thinking starts with an analysis process that focuses on fact or truth finding. When we make a decision, our behavior is based on the facts or on scientific truth. We need the facts in order to make our decisions. Therefore, we need objects to analyze in order to find the facts and the truth. This approach is a Scientific-Research-Analytical-Fact-Finding one.

Breakthrough Thinking starts from the ideal (substance or essence). When we make a decision, we base our behavior on the purpose, which is the ideal (substance or essence). This approach focuses on Design-Purpose-Solution finding. For example, because of the increasing competitive pressures of rapid change, Toyota found it necessary to redefine the ideal (substance) of its products when it looked to the future. Therefore, Toyota established the "Research Institute for Substance" because it realized that it could not find the substance for a car by trying to analyze the engine of the car.

The three thinking paradigms are completely different and each has a different approach. We cannot neglect any of these three thinking paradigms because each has an influence in the decision-making process. We have to select and utilize each of these paradigms on a case-by-case basis. Someone who uses and interchanges these thinking paradigms is referred to as a "Multi-Thinker."

Why Multi-Thinking Paradigms Are Needed

"The fact that solution paradigms shift continuously — and succeed one another with increasing rapidity and impact — should serve to make clear that the future cannot exist along the same lines as the past and present. Yet we persist in depending on the demonstrably insufficient methods of quantitative data collection, analysis, and general prediction that are derived from the pervasive influence of Conventional Thinking. It's "Push Thinking" from the past."[5] If we agree with the assumption that the future cannot exist along the same lines as the past and present, then we cannot use push thinking techniques from the past. Therefore, fact-finding approaches for finding solutions do not work well.

We need the opposite approach from push thinking. We call it "pull thinking" which learns from the future. Pull thinking is thought of as a design approach. Since we can not predict the future by learning from the past and present, we should design our future and learn from what we designed.

When we design the future, we have two approaches. One is the push-style design approach based on the past and present facts utilizing conventional thinking. The other is the pull-style design approach based on the purpose or substance of things, and utilizing Breakthrough Thinking. It is clear that the latter approach is effective because it works on the assumption that the future cannot exist along the same lines as the past and present. We need Breakthrough Thinking to facilitate finding future-oriented solutions.

Changing Our Perspective — From Machine View to System View

Since there is no future that continues along the same lines as our past and present (because of the drastic changes going on in the world), we cannot find futuristic solutions based on past and present facts. Our thinking base should be changed away from facts and refocused on the substance (essence or ideal) of things.

How do we identify the substance of things? It is not easy for us to find this substance. We have to transform ourselves from having a conventional machine view to having a systems-oriented view. The traditional perspective of conventional thinking is to view things as a reductionistic machine, breaking everything down into elemental parts, and neglecting the "whole" organic view.

The epistemology of Breakthrough Thinking is that "everything is a system," which is defined as follows:

1. Every system has elements.
2. Each element is interdependent or interrelated. We cannot divide out elements, like the machine perspective.
3. Every system has purpose(s) or function(s).
4. We need to take a holonic view.

By this definition, we can see everything as a system. For example, "love" is a system. When we love somebody else, we love each other. I love you. You love me, which is interrelated. If you cut this interrelationship, love will disappear. The purpose of love is "to enjoy love" or "to satisfy basic human desires." Love is a part of my life. Since these statements satisfy the four basic definitions of a "system," we can recognize that love is a "system."

Another important point of the system view is the "holonic view." If we define everything as a system, then everything is a "Chinese box," which means that a bigger box (system) includes a series of smaller boxes (systems). A small box (system) contains still smaller boxes (systems), and so on. Each box (system) has its purpose(s). If you repeatedly ask "What is the purpose?" and then "What is the purpose of that purpose?" and then "What is the purpose of that purpose of that purpose?" etc., you can reach the biggest box, which is "wholeness." You can view everything from the perspective of this wholeness. This is how we would identify and focus on this wholeness. We start by asking for the purposes.

The purpose is the substance of all things because everything is a system and every system has a purpose. Asking the purpose of the purpose is a pursuit for the substance (wholeness) of the thing (system). We call this search the "purpose expansion."

If you can have the epistemology of the systems view, then you can tackle the problem by using purpose expansion, not by the collection and analysis of past and present facts. It is for this reason that Breakthrough Thinking in the initial stages focuses on the purposes. Without an understanding of the epistemology of the systems view, you cannot understand this new thinking paradigm. For a deeper understanding of this process,

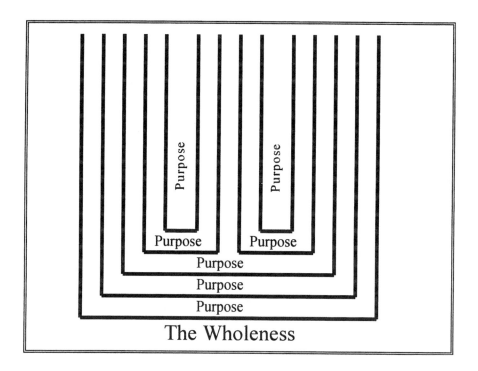

you should read the two books listed earlier, *Breakthrough Thinking* and *Creative Solution Finding*.

Breakthrough Thinking (BT) creativity won the Showa Ceramic Company of Seto City, Japan, the 1993 Nikkei Advertising Award. Showa management has been using BT processes for more than 20 years, and utilized when to relieve its increased demand for tile samples which had overloaded its capacity. The three employees responsible for generating and distributing the samples were working long hours, and when one of the employees had a long-term illness, the demand for tiles had become impossible to satisfy. A BT meeting was held to discuss the situation. Without considering the past or the current situation, they immediately focused on a purpose hierarchy that opened their eyes to innovative solutions. The purpose hierarchy went as follows:[6]

1. Send the wall tiles as a sample
2. Send a model of the wall tile
3. Send a representation of the wall tile, color, and texture
4. Send a representation of the tile's color and texture
5. Send the beauty and feel of the wall tile.

During the purpose expansion they realized that what was wanted was not an actual tile sample, but rather a representation of the shade and texture of the real tile. With this understanding they proceeded to develop the focus purpose using the solution-after-next principle and they identified an ideal target. As a result of this BT process, they now send out a small piece of the original tile, referred to as a "stick sample."

The "stick sample" has conceptually changed the advertising process through samples. A total cost reduction of 70% was achieved. Production costs are reduced, materials costs are down, mailing costs are reduced because this new sample can be sent through the regular mails, and productivity has increased. One person is now able to handle the entire demand, including the increases in demand, for sample tiles.

By ignoring the "problem" (past and present) and looking for the ideal solution ("What are we really trying to accomplish?") through purpose expansion, Showa Tile Company was able to meet its demand for tile samples with reduced costs and fewer employees. This BT innovativeness won it the coveted Nikkei Advertising Award.

The Breakthrough Thinking Paradigm

Breakthrough Thinking consists of a thinking paradigm and thinking process. The thinking paradigm of Breakthrough Thinking is the opposite of the paradigm of conventional thinking. Its main points are expressed as seven principles.

1. Uniqueness principle
2. Purposes principle
3. Solution-after-next principle
4. Systems principle
5. Needed information collection principle
6. People design principle
7. Betterment timeline principle

Uniqueness Principle

Assume initially that the problem, opportunity, or issue is different. Don't copy a solution or use a technique from elsewhere just because the situation may appear to be similar. Even if your family and your friend's family are similar, the combination of DNA of each family member is completely different. The uniqueness principle is based on the following reasons:

Creative (Breakthrough) Thinking **17**

1. No two situations are alike. At the very least, the people in each are always different.
2. Each problem or opportunity or issue is embedded in a unique array of related problems, opportunities, or issues.
3. Problems, opportunities, or issues that may look alike may have different purpose needs.
4. Tomorrow's technology is already different from today's.

In using this principle, we have to think about the locus or solution space of the problem. This locus is defined using three points.

1. Who are the major stakeholders? Who's viewpoint is most important? For example, if you would like to design a bonus system, who's viewpoint is important? The purpose of a bonus system is different from the perspective of management as compared to the labor union.
2. What is the location? In Japan? In California? In India? In the factory? In my family? Solutions and purposes differ in each of these locations.
3. When (what is the timing)? In the year 2010? Next year? If we have a longer term solution space, our solution is bigger than the short-term solution.

This uniqueness principle changes the general solution into SSS (situation-specific solutions) for the specific locus. We should request situation-specific solutions, and not focus on generalized truths or facts. Conventional thinking focuses on generalization and similarity, whereas Breakthrough Thinking focuses on situation-specific solutions and uniqueness.

Purposes Principle

Explore and expand purposes in order to understand what really needs to be accomplished and to identify the substance of things. You can tackle any problem, opportunity, or issue by expanding purposes, if you change your epistemology into a systems view. Always start any activity with a hierarchy of purposes like a Chinese box. For example, just look at your pencil.

What is the purpose of your pencil?
  The purpose of a pencil is to write letters.
What is the purpose of writing letters?
  The purpose of writing letters is to display letters.
What is the purpose of displaying letters?

The purpose of displaying letters is to show information.
What is the purpose of showing information?
The purpose of showing information is to transmit information.
What is the purpose of transmitting information?
The purpose of transmitting information is to transfer information.
What is the purpose of transferring information?
The purpose of transferring information is to transfer my intention.
What is the purpose of transferring my intention?
and so on.

One of unique features of Breakthrough Thinking, when compared with conventional thinking, is the expansion of purposes from a small purpose through a hierarchy of larger purposes. Understanding the context of purposes provides following strategic advantages:

1. **Pursue the substance of things.** We can identify the most essential focus purpose or the greater purpose, often referred to as the substance (core element) of things by expanding purposes.
2. **Work on the right problem or purpose.** W. Edwards Deming said that *"It is not enough to just do your best or work hard. You must know what to work on."*[7] From this we see that knowing clearly what to work on is the major benefit of the purpose principle. Focusing on right purposes helps strip away nonessential aspects to avoid working on just the visible problem or symptom.
3. **Improve the ability to redefine.** Redefining is usually very difficult. For example, a pencil is for writing letters. We will experiment with redefining this pencil. It is easy to redefine a pencil if we have its purpose hierarchy.

    A pencil is not to write letters, but to display letters.
    A pencil is not to display letters, but to show information.
    A pencil is not to show information, but to transmit information.
    And so on.

    Once you've redefined the pencil, you can have different viewpoints, each of which enables you to solve problems from different directions. Breakthrough Thinking can improve your ability to redefine problems or opportunities, and to identify creative solutions.
4. **Eliminate purpose/function.** From Systems Theory we learn that a bigger purpose may eliminate a smaller purpose. For example, if you just focus on the purpose of transferring information, then you don't need to write letters because you can utilize the telephone for

the focus purpose. Sometimes, the elimination of current work, or of a system, or of parts of a system, is the best solution. By focusing on the bigger purpose, you can eliminate unnecessary work/systems/parts, which means that you can get more effective solutions.
5. **More options, more creative.** If you have a purpose hierarchy, you have a lot of alternative solutions. For example, you may imagine a pencil, ball-point pen, chalk, writing brush, etc., are all options for writing letters. You may use a pencil, ball-point pen, chalk, writing brush, overhead projector, etc., for displaying letters. You may consider a pencil, ball-point pen, chalk, writing brush, overhead projector, video, movie, telepathy, telephone, Internet, fax, etc., for transferring information, and so on. A bigger purpose creates more solutions or options. You can have a full range of options for a series of purposes. This opens the door for increased creativity.
6. **Holonic view.** If you instruct your subordinate to make a copy for you with instructions such as "Please copy this paper," your subordinate copies it without thinking. If you give instructions such as "Please copy this paper for the purpose of presenting it at a large hall tomorrow," your subordinate will think it is for a presentation and will make an overhead transparency sheet and expand (blow up) the size of the sheet. In the second case you wouldn't need to give further instructions to have the transparency created. If you have a hierarchy of purpose, you may think or talk or behave from the bigger holonic view and you will thereby improve your productivity.

Solution-After-Next (SAN) Principle

Think and design futuristic solutions for the focus purpose and then work backward. Consider the solution you would have recommended if in 3 years you had to start all over. Make changes today based on what might be the solution of the future. Learn from the futuristic ideal solution for the focus purpose (pull thinking) and don't try learning from the past and present situation (push thinking).

The root of the SAN process is its focus on the purpose or the substance of things, not based on the present and past situation. In conventional thinking, the visions or objectives are based on the present and past. For example, if the profits for last year were $10,000,000, and if this year is more profitable, then our objective for next year is a 15% increase in profit. This scenario is push-style thinking.

In conjunction with the purpose principle, we need more options for the situation-specific-solutions. We have to always seek for more options; for better solutions. The best solution will come from many optional ideas, not just one great idea. "More options brings more creativity" is a basic principle of the creative process.

Conventional thinking seeks for the ultimate one right answer (truth). Breakthrough Thinking seeks for many options (solutions) that will satisfy the focus purpose or substance of things.

Systems Principle

Everything we seek to create and restructure is a system. This is the systems view epistemology of Breakthrough Thinking. Think of solutions and ideas as a system. Use a solution framework that includes all elements and interrelationships. The solution-after-next or visions or objectives are usually not clearly defined or are possibly even unseen. A poorly detailed solution creates multiple problems to take the place of the original situation. Understanding all elements and the interrelationships of the SAN (visions or solutions) allows you to determine in advance the complexities you must consider in the implementation of the solution. The system principle of Breakthrough Thinking defines these elements and their interrelationships. In most cases, this principle helps in implementing the solution-after-next target (concept or dream).

When you see everything as a system, you have to consider the eight elements of a system in order to identify the solution.

1. Purpose: mission, aim, need
2. Input: people, things, information
3. Output: people, things, information
4. Operating steps: process and conversion tasks
5. Environment: physical and organizational
6. Human enablers: people, responsibilities, skills, to help in the operating steps
7. Physical enablers: equipment, facilities, materials to use in the operating steps
8. Information enablers: knowledge, instructions

Then you consider the six dimensions or characteristics of each element of the system.

1. Fundamental: seven elements as a fundamental base: what, how, when, why, where, who, how much.
2. Values: motivating belief, global desires: increase profits, protect the environment.
3. Measures: objectives, goals, performance specifications, such as How much profit? How well do we perform in environmental protection?
4. Control: how to evaluate and modify elements of the system or the entire system as it is operating, such as how to get things back on track, how to take corrective action.
5. Interface: relationship of all systems dimensions to other systems or other elements, such as What other systems or elements will be affected by this one? What impact or friction will this have on other systems? How to decrease internal and external fluctuations? How to have the network synergy benefit other systems, the company as a whole, a group, or the organization?
6. Future: planned/changes, such as What is the next solution after this solution? How should we prepare for the next change?

As an example, let's consider a "Love System":

What is the purpose of a love system?
  To enjoy love.
What is the input?
  Lack of enjoyment by the man and the woman.
What is the output?
  Enjoyment by the man and the woman.
What are the operating steps?
  Meet at home, go to a restaurant, have lunch together, go to a movie, etc.
What is the environment?
  Romantic, friendly, relaxed, etc.
What are the human enablers?
  The hostess, lunch cook, movie theater owner, etc.
What are the physical enablers?
  Sports car, drink cups, lunch dishes, movie projector, etc.
What are the information enablers?
  Lunch menu, movie film, etc.
What are the values for each element?
What are the measures for each element?
What are the controls for each element?
What is the interface of each element?
What is the future for each element?

We have to ask these questions continuously until we can understand the contents. Once we can understand the contents of the ideal targets, the possibility of implementation and the opportunity for success increase greatly. We can make our innovation happen by using systems principles.

Needed Information Collection Principle

Collect only the information that is necessary to continue the solution-finding process. Know your purposes for collecting data and/or information. Study the solutions, not the problems.

In Descartes thinking, we cannot think without the facts, so we collect data and information and analyze it. We can find the problems in a specific area and we become an expert about the problem. We waste an enormous amount of time collecting the wrong information, analyzing it, compiling it, storing it, and identifying the wrong problems in our so-called information society.

In Breakthrough Thinking, we try to collect only the needed information, utilizing the following guide:

1. Determine what purposes would be achieved by collecting the information before taking the time and effort to collect and analyze the information.
2. After you know your purposes and your target concepts, you should start collecting the needed data and information, focusing on implementing your target concepts or solutions.
3. Collect information on purposes and possible solutions rather than on the problem — extensive information collection makes you an expert on the problem but it may prevent you from seeing very many alternatives or solutions.
4. Ask how the solution could be improved on or would work differently by recollecting the information after a period of 3 months or after 1 year.
5. Collect information from a variety of sources and share it broadly.
6. Don't predict the future by collecting data of the past and present. The future cannot be predicted from a perfect knowledge of the past and present, because the future does not follow the same line as the past.
7. Focus on relevance. Relevance is better than accuracy.
8. Don't trust data and information. Data and/or information is only a representation of the real world and collecting information is not a neutral process. The collection often distorts the results.

The benefits of limiting data and information collection are

1. It requires less total time, cost, energy, and document handling.
2. It provides a high quality of usable information.
3. It eliminates the defensiveness of people when the data is being reviewed.
4. It makes it possible to identify better system interrelationships.
5. It makes it possible to provide guidance to implementing solutions.
6. Coming to know that information and knowledge are not power, knowing how to use information and knowledge is power.
7. It improves thinking productivity drastically.

People Design Principle

Give everyone who will be affected by the solution or idea the opportunity to participate throughout the process of its development. A solution will work only if people know about it and help to develop and improve it.

> No matter how wonderful an idea is, if the employees don't support it, it will die.
> No matter how poor an idea is, if the employees have ownership in the idea, it will succeed.

There are two major reasons for using this principle. One is that individuals are the core to a solution's success. The implementation of planning means that people's values and behavior must be changed throughout the process of the project. According to the Participant Handbook for Breakthrough Thinking:

> No matter how well a system or solution is conceived, designed and executed, if people don't like it, it will fail. No matter how poorly a system or solution is conceived, designed or executed if people want it to work, it will succeed. The value people see in a system or solution will significantly impact the results. People can only become aware of the reality of an idea by interacting with it, by creating different possibilities through their own processes of thinking. Bringing people actively into design and problem solving is a need, not just a desirable social

value. Getting people to participate not only gives you access to their ideas, but makes them partners in the enterprise, increasing your chance of success.[8]

With this principle "we try to build the collective genius team, which means the team becomes genius as a group, but each individual is not a genius."[8] Everyone has the potential to be a valuable contributor. Individuals with different kinds of knowledge and experience in planning and solution finding should be consulted — cross-functional teaming. Their concerns and ideas must be treated as the fabric of excellent solution finding. People enjoy working on and accepting responsibility for a project, especially when they understand the principles of uniqueness, purpose, solution-after-next, systems, etc. Then they will be more effective and productive, and as a result the solution will be more successful.

Betterment Timeline Principle

Install changes with built-in seeds of future change. Know when to fix it before it breaks. Know when to change it.

Rapid change creates problems. No matter how effectively you can create solutions or improve them, it will never be good enough to meet the needs of the future. People and environments will change. New technology will render certain processes obsolete quickly. The original assumptions and conditions will go away. Everything is facing the big change. No solution is ultimate. The solution is change itself.

The Betterment Timeline Principle makes solutions effective for the future. The solutions regularly have to be changed and upgraded toward the solution-after-next target. The assumptions, purposes, values, measures, technology, environment, and constraints of the solutions should constantly be changed over time.

We should ask the following questions when we have defined a solution.[9]

1. Do we think there is only one ultimate solution and this is it?
2. Do we have plans to "fix it before it breaks"?
3. Have we prepared a schedule for initiating change and improvement?
4. Does our solution include the seeds of its own future change (built-in change)?
5. Have we designed our strategic and financial plans with the needed betterments integrally included?

Creative (Breakthrough) Thinking 25

The solutions in our society require the dynamic consistency of change. Consistently seek the solution-after-next and the solution-after-solution-after-next and so on. You must design a new set of rules for the future. It may appear that many organizational rules are being discarded or broken; however, Breakthrough Thinking is a way for you to learn how to continually design permanent rules effectively.

Breakthrough Thinking Process

The thinking processes of conventional thinking and Breakthrough Thinking are very different. The thinking processes of conventional thinking include

1. Collect data.
2. Analyze data.
3. Find out what's wrong. Identify facts or truths.
4. Replace it by an alternative.
5. Find new solutions.

Conventional thinking is a problem (truth, fact) finding process and uses the analytical (or research) approach. On the other hand, Breakthrough Thinking is an approach of reasoning toward a situation-specific solution and uses a design approach. It is an iterative, simultaneous process of mental responses based on the Breakthrough Thinking Seven Principles, called the Purpose-Target-Results Approach (PTR Approach). PTR's three phases are

1. Purpose: Identifying the right solution by finding focus purposes, values, measures
2. Target: Targeting the solution of tomorrow — ideal SAN vision and target solution
3. Result Getting and maintaining results toward implementation and systematization

In these three phases, you don't need to apply all seven principles all the time, but become aware and remain aware of all seven principles as you search for solutions. For examples, you have to think uniqueness, purpose, betterment timeline, people design principle for people involvement, etc. in each phase. In each phase, there are three substeps referred to as the list-build-select (LBS) steps. The three LBS steps are

1. List: List many alternatives.
2. Build: Build options or solutions, by integrating alternatives.
3. Select: Select a workable option or solution.

From the PTR phases and the LBS steps we get a matrix shown below:

	List	Build	Select
Purpose	Many purposes	A purpose hierarchy	Focus purpose(s) measures of success
Target	Many ideal and possible ideas or alternatives	Alternative options or solutions	SAN target
Result	All elements and related systems	Systems details using the Systems Matrix	The "first release" and plan for betterment

Tokyo Power Company, an electricity producer, was having trouble with its power towers. Crows living in Japan found the towers to be excellent locations for their nests. Unfortunately, they would often use conductive substances such as wires to build their nests, causing shorts, power outages, and serious damage to the lines. One year, the crows managed to put the bullet train out of commission for 4 hours during a time of heavy snowfall.

Frustrated by the crows, the Tokyo Power Company, in an attempt to solve the problem, established the Bird Prevention Institute and funded it with $3 million U.S. The institute was given 3 years to research the crow problem. Researchers studied the crow, its characteristics and habits. They tried balloons, various frequencies of noise, etc., all of which worked for a short time, but eventually the crows would return back to their old habits. They found the balloons to be entertaining toys. After 3 years, the Bird Prevention Program was disbanded.

One day, in a conversation with Mr. Seki of the Hitachi Wire Company, the crow problem was brought up. Mr. Seki is trained in Breakthrough Thinking and suggested that they try a new approach to the problem. Rather than using conventional thinking to study the past and present status of the crow, its characteristics and its habits, he suggested doing a purpose expansion. He asked, "What is the purpose of the Bird Prevention Institute?" Then he asked what the purpose of that purpose was, and so on. Within about 20 minutes, this purpose expansion exercise brought the executives to the realization that what they wanted was not to get rid of the crow, but to keep it from causing power outages. They decided that perhaps they should be looking at ways to welcome the crow and learn how to live together. After another 5 minutes of discussion, they decided that if they were to construct artificial nests for the towers that would not cause shorts, then the crows would have a place to stay, and the damage caused by the crows' placing inappropriate materials on the power lines would be ended. They decided to build the artificial nests and test them for 1 year. During that year, damage from the crows went down to zero. This process of Breakthrough Thinking (Concept Creation) had accomplished more in 25 minutes than the entire institute was able to accomplish in 3 years.

A World Class Manager is a Breakthrough Thinker

Breakthrough Thinking was born from the research of successful people. World Class Managers should be Breakthrough Thinkers because they should be successful people at a world-class level. World Class Managers need to focus on purposes, not on causes. They seek for uniqueness, not similarities. They are not problem experts, continuously fighting fires, but they should be solution experts. They gather only the needed information, not all information. They love soft data, not hard data. They try to involve people from a broad area, not from only the problem area. They can treat ambiguous conditions without firm scope. They always make decisions based on the solution-after-next principle using strategic vision. They have a systems view and think in terms of both wholeness and interrelationship. They use the PTR approach, not the analytical approach. Now that you understand that World Class Managers have a different thinking paradigm, we can further define World Class Managers.

Only if you're different do you make a contribution,
Otherwise you're just an imitation.

Endnotes

1. Albert Einstein, "Telegram to Prominent Americans," *The New York Times*, May 25, 1946.
2. Gerald Nadler and Shozo Hibino, *Breakthrough Thinking*, Prima Publishing, Rocklin, CA, 1995.
3. Gerald Nadler and Shozo Hibino, *Creative Solution Finding*, Prima Publishing, Rocklin, CA, 1995.
4. The "Paradigm Shift in Thinking," "Breakthrough Thinking," and most of the other catch phrases in this chapter are registered trademarks of *The Center for Breakthrough Thinking, Inc.*, which is housed at the University of Southern California (USC) in Los Angeles, CA.
5. Nadler and Hibino, *Creative Solution Finding, p. 85.*
6. *The Breakthrough Thinker*, The Center for Breakthrough Thinking, Inc., Summer 1993, Los Angeles, p. 2.
7. W., Edwards Deming, *Out of the Crisis*, MIT Center for Advanced Engineering Study, Cambridge, MA, 1986.
8. "Breakthrough Thinking," *Participant Handbook*, The Center for Breakthrough Thinking, p. 18.
9. Ibid., p. 318.

Chapter 3
World Class Management

The greatest wasted resource is the human resource.
Walt Disney

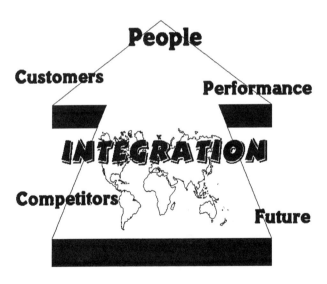

Don't let any system control you, whether it's the computer system, or the production control system, or the accounting system, or your data collection system. The system is yours to use, not yours to be controlled by.
Gerhard Plenert

My family and I lived in Malaysia for one school year.[1] We enjoyed integrating into the culture and lived in university housing, which, unlike the housing we were used to in the United States, had no screens on doors and windows. The kitchen was connected to a laundry room, and the laundry room had open holes to the outside. This laundry room became a tourist attraction for the local frog population, and we had regular visitors. One day, one visitor made its way into the kitchen. The children, in their excitement, tried to catch the frog. They boxed it off against one corner of the room and blocked off the other two sides with their hands. They were sure that the frog couldn't get away, being boxed in from all four sides. But this frog didn't think like the American frogs they were used too. It jumped high and attached itself to the side of the wall. Then with its next jump, it safely escaped its barrier, much to the surprise of the children. This frog had behaved like a monkey. It took on an unexpected perspective and beat its competition. That's what World Class Management is all about. It's about taking a different, unexpected, competitive perspective that will surprise and bypass the competition.

World Class Management is broad in its application, and numerous publications discuss the subject in detail. However, in order to get a clearer understanding of how World Class Managers manage change, we will discuss

1. People
2. Customers
3. Performance
4. Competitors
5. Future
6. Integration

People — The Source of Change

> *Thinking is the hardest work there is, which is the probable reason why so few people engage in it.*
>
> **Henry Ford**

Federal Express uses a pyramid to describe its goal focus as an organization. At the top of the triangle is People, on the bottom left is Service, and on the bottom right is Profit.[2] In the words of Frederick W. Smith, Chairman and CEO of FedEx, "Federal Express, from its inception, has put its people first both because it is right to do so and because it

is good business as well. Our corporate philosophy is succinctly stated: People-Service-Profit (P-S-P)." For FedEx, People equals Employees. Taken from FedEx's Manager's Guide we read, "Take care of our people; they, in turn, will deliver the impeccable service demanded by our customers who will reward us with the profitability necessary to secure our future. People-Service-Profit, these three words are the very foundation of Federal Express."

Why are employees so important in the change process? Because they are the source of all innovative, creative, positive change! It becomes extremely important to make employees feel as though they are a part of the overall success of the organization. This is accomplished through a series of steps. The first step is to recognize that the purpose of a measurement (performance evaluation) system is neither for record keeping, nor for data collection. The purpose of a measurement system is to clearly define direction for the employees, thereby motivating them in the desired direction.

A few years ago I was asked to visit a factory targeted for closure.[3] The plant had what was considered to be one of the best quality improvement systems available, with quality circles, statistical process control, slogans, banners, training, and motivational programs. Yet the plant was being closed because of poor quality products and the resultant loss of customers. The quality manager asked me to evaluate "what has gone wrong"; the managers didn't want to make the same mistakes at their next place of employment. In discussing their quality environment, I asked how employee performance was measured. They said that employees were measured by units produced, compared against standard number of units of output. Employees were paid a performance incentive pay based on the number of units produced. I asked them if employees were rewarded for good performance. The managers told me about awards, banners, and certificates. I asked if employees would be motivated more by a certificate or by an increased paycheck. They agreed that pay would be the greater motivator. Then I asked: "Why would employees be interested in improving quality when the rewards (a greater paycheck) were in working fast without regard for quality?"

Often, when I discuss the measurement/motivation relationship, individuals will ask; "Why don't we just measure everything, including what we want to motivate?"

The answer is simple. Measuring everything is as good as measuring nothing. What we are trying to do is motivate a response. Measuring everything confuses the issue and doesn't help motivate anything. Therefore, the key to understanding what to measure can be stated as follows: "The purpose of a measurement (evaluation) system is motivation, not data collection. Measure exactly and only what you want to motivate."

The measurement–motivation relationship clearly defines for the employees what is important in their job functions. This relationship defines goals, and without a clear-cut measurement system, it would be impossible to achieve this goal.

The concept of measurement and motivation goes hand in hand with another employee change-focused concept: team-based empowerment and gainsharing. Teaming is the key to synergistic results. One person's ideas trigger another's ideas, and the chain reaction of ideas generates unending creativity; however, teaming is an art and should not be confused with "grouping." Grouping is bunching people together in a room and holding a discussion. Teaming is a focused, long-term developed creativity, which becomes removed from the politics and status-grabbing problems of grouping. Teaming is the removal of fear — the fear of criticism, or the fear of having ideas stolen. Teaming takes time and requires direction and, as discussed under measurement and motivation, teaming requires goal-focused direction in the form of a charter.

Teams, properly constructed with charters (goals and procedures), empowerment, and gainsharing, develop a sense of ownership with team members. Ownership means that employees no longer see change and the change process as something decreed from above. Rather they see themselves as in control of the changes and the change process, deciding what should be changed and how. Employees no longer see themselves in the traditional perspective where their purpose is solely to keep their jobs and get a raise. They see how they fit into the larger needs of the organizational structure. Chapter 4 specifies the objectives of the team, what they are empowered to do, and how they will be rewarded for their performance (gainsharing). This is a critical element of Total Quality Management (TQM).[4]

> *Treat a man as he is and he will remain as he is.*
> *Treat a man as he can and should be and he will become as he can and should be.*
>
> **Goethe**

A second step toward making the employees feel a part of the organization is for them to work in an ethical, respectful environment. If your employees find themselves in a working environment that focuses on fairness and integrity in relation to customers and vendors, then they will also feel the need to perform in a similar manner. Additionally, the employees are likely to expect the same fairness and honesty in employee–manager relationships. For example, it's one thing for FedEx to claim that employees are the first priority, but this needs to be demonstrated via corporate responsiveness to employee concerns. FedEx responded to these concerns by installing a program in which employees evaluate the performance of their managers, and a poor report from employees can result in the dismissal of the manager. Additionally, FedEx has installed the Guaranteed Fair Treatment Procedure (GFTP), which "affirms an employee's right to appeal any eligible issue through a systematic review by progressively higher levels of management."[5]

If a company builds its reputation on value-based ethical behavior, employees are likely to believe them when they say, "We want your ideas, even if they seem crazy or critical. All ideas are good ideas and will be carefully considered!"

In the movie *Quiz Show*, a film based on a true story, an academic intellectual gets caught up in a scheme in which he is the contestant on a television game show. In this show he is expected to answer tough questions about a variety of subjects, but he is fed the questions and their answers ahead of time so that he is set up as the winner. The objective was to increase viewer interest in the show. This scheme was successful for a time, but then the scheme was discovered, and the participants were questioned by a national inquiry board.

The father of the contestant was a man of high reputation. The inquiry made national headlines and caused the father a great deal of embarassment. The son could lie and say he was not part of the scheme, and his name, and the good name of his father, would be cleared of embarrassment. This would be easy, since all other participants of the scheme who had testified before him had already maintained the lie. However, he felt that the lie had gone far enough. He went to his father, told him the truth, and asked for his forgiveness, then he asked his father what he should do. His father, a man of great integrity, was less worried about his reputation than he was about his son's character. He told his son that he would support him at the inquiry, and he did. Here was a man (the father) who thought more of someone else (his son) than about his own hard-earned reputation. His son had ruined his reputation, yet he forgave him and immediately stepped forward to help. Would we be as forgiving of our family members? Would we be as forgiving and patient with our employees? It reminds me of the saying, "We want others to judge us by our intentions, but we judge others by their actions." Are we as forgiving of others as we want them to be of us? Effective managers build trust in employees by being understanding and forgiving.

If you don't have the trust and respect of your employees, they're not "your" employees in spirit. Support of the employees and their ownership of the change process are critical to World Class success. Your fairness, from the perspective of the employees, is critical. Keep them informed and involved, so that they understand and empathize with your concern for them and your commitment to fair treatment. *Never* betray that trust, because, as Ricardo Semler of Semco says, "Fairness is for employees like quality is for customers — it takes years to build up but collapses over a single incident."[6]

Once, when my family and I were going to a movie together, we thought we'd stop at the store to buy snacks to take into the theater with us — the treats at the show are always so expensive, we do this to save cost. We would then smuggle the treats into the show in our pockets or in a purse. As we were standing in line, Zack, my eight-year-old son, blurted out loudly to Mom, "Are we going to stick the candy in our pockets?" That sounded somewhat questionable and caused people in the store to look at us rather strangely. But how could we be mad at him? He was just asking what we were doing. He assumed this to be a reasonable question, in light of past behavior. If we didn't like the embarrassment, we shouldn't have taught him this inappropriate behavior. This was a lesson about honesty; what we considered a minor indiscretion was teaching our children a false principle. Similarly, in our own companies, if we don't behave honestly toward vendors, customers, or employees, let's not be surprised when we get embarrassed.

> Your actions speak so loudly I can't hear a thing you're saying.

Another key to developing a change-focused workforce is found in training and education. In a special report by *Business Week* on "The New Factory Worker," we find statements like "Today, life on the line requires more brains than brawn — so laborers are heading for the classroom." "Factories are training more and hiring better-educated workers."[7] The more employees understand your management, production, accounting, etc., processes within your organization, the better equipped they are to change (improve) it. Additionally, the more new ideas they are given, the broader their perspective and their idea base become. Employees should visit the sites of customers, competitors, and vendors; they should attend conferences and exhibitions; and they should subscribe to formal education programs, both internal and external. The broader their base of knowledge, the better their ideas become.

Making employees the source of change within your organization should be the number one priority in a change-oriented organization. As the master motivator Stephen R. Covey says in his book *The Seven Habits of Highly Effective People*, "always treat your employees exactly as you want them to treat your best customers".[8]

Nationwide employee cutbacks from December 1992 to September 1995 yield the following numbers:[9]

Company	Cumulative Laid-Off Workers	CEO Annual Compensation
IBM	122,000	$4,600,000
AT&T	83,000	$3,499,000
GM	74,000	$3,425,000
Boeing	61,000	$1,445,000

What message does it give our employees if, every time there is a problem, we attack them first? Is this positive, growth-oriented change, or desperation? Are we saving our organizations, or destroying the root source of what makes our companies successful — are we destroying our core competencies? Is the CEO of these companies truly as valuable as 100 employees? Or is the wrong person getting laid off? Is it any wonder that the goals of our employees are to "keep their jobs and get a raise?" When we start treating employees as customers, as FedEx stresses, then they will start responding accordingly, as the Goethe quote stressed.

Customers — The Reason for Change

Customers define the success of our organization. An occasional customer is good only if we are satisfied with being an occasional organization. A repeat customer defines success, and a repeat customer requires responsiveness. Traditionally, the best World Class definition of customer satisfaction (customer quality) has been "a Satisfied Customer is one who is so excited about your product or service that they wouldn't think about purchasing it from anyone but you."

Defining customer needs (getting customers excited) is more than simply asking them what they want. Often they do not have a clear definition of what they want, because they don't know what their options are. Customer satisfaction often requires helping them define what is best for them. This requires getting into the customer's heart and soul; it requires becoming a friend and caring about them personally.

AT&T uses a measure called Customer Value-Added, which measures and compares how satisfied AT&T customers are with AT&T service, vs. AT&T competitors' customers' satisfaction with their competitors' service.[10] This type of measure motivates a higher level of customer responsiveness than does some loose definition of "satisfied customers," which has no measurement/motivation system to encourage a change in behavior.

Another good way to evaluate customer satisfaction is to actually become a customer. Purchase your own products and services and see if you are satisfied, and then compare your own performance with your competitors' by purchasing their products. For example, nothing teaches an engineer more about product usability or manufacturability than making the engineer become his or her own customer by having them go into the factory and manufacture the product from their own designs, and then use the product themselves. Similarly, nothing teaches accounting staff members more about the usability of the forms and procedures they develop than making them use the forms and follow the procedures for a few weeks themselves. It's amazing what types of redesigns this process can generate!

It is important to point out here that the definition of customer satisfaction given above is based on conventional thinking, and that an entirely new, enhanced perspective on customer satisfaction is generated through the synergy of Concept Management. This new, synergistic definition goes miles beyond the World Class definition given earlier — the new focus is on delighting, and not just satisfying, the customer. I'm sure you'll be anxious to read Chapter 5 to learn more.

Please note that I haven't drawn a large distinction between internal and external customers. Ultimately, the only customer that matters is the external customer. Sometimes, satisfying the internal customer requires unnecessary and wasteful bureaucracy, and this customer is better eliminated (eliminate the function, never the person) than satisfied. However, we have already discussed how all employees should be treated as customers, and by this we mean treat the employee like a human being, but attack any and all processes as opportunities for change.

Performance — Measures and Motivators for Change

There are a multitude of performance measures, and we will discuss a few. However, in order to stress the point made earlier, the purpose of these measures is motivation — NOTHING ELSE! Some of the most common measures include financial performance measures and operational performance measures. For example,

> Financial Performance Measures
>   Profits
>   Return on Net Assets (RONA)
>   Sales
>   Financial Ratios
> Operational Performance Measures
>   Quality
>   Productivity
>   Efficiency
>   Throughput (quality units delivered to the customer)
>   Inventory Levels
>   Value Adding
>   Benchmarking

Financial Measures are anti-World Class, for numerous reasons. First off, because they are short-sighted and short-term oriented; they motivate short-term thinking. For example, the Saturn plant is an overwhelming operational success story at General Motors. Its quality is impressive, and the customer satisfaction is outstanding. So why don't we blow up all the old-style, outdated GM plants and replace them with Saturn technology? The reason is simple. To make such a dramatic change, the company would have to go through major restructuring, running in the red for years (corporate losses) — and no CEO could survive the turmoil caused by such a revamp. Since the objective of all employees is to "keep their job and get a raise" (even for the CEO), by satisfying their measurement system, the CEO wouldn't dare do what is logical and best for the long-term survivability of the company. The CEO is more interested in short-term survivability — his survivability.

Second, short-term financial measures are internally conflicting and therefore frustrate themselves. Everyone is familiar with how businesses spend the first 90–95% of the period working toward profitability goals, and then spend the last few days of the measurement period rushing around, throwing profitability to the wind in an attempt to meet a sales goal. At what point will we finally realize that the economic production function was right after all, and that we can't maximize profitability and maximize sales simultaneously? Something's got to give, and it happens when the business is run two different ways during the two different parts of the period.

Third, financial measures defy logic by defeating operational efficiency. Everyone has heard stories such as the plant manager who thought he would improve his operational efficiency by eliminating half of his inventory. After all, everyone knows that excessive inventory is a "waste" and costs financing dollars, making a company less competitive. However, at the end of the period, after dropping inventory by 50%, one of his key performance measures was the current ratio, a measure of current assets against current liabilities, which supposedly demonstrates an ability to pay one's debts. By dropping his inventory, a current asset, he had nearly destroyed his current ratio. He was told to "fix the problem by the end of the next period or get fired." Since his objective was "to keep his job and get a raise," he responded to his measurement system, and during the next period he purchased his inventory back. Now he has wonderful current ratios.

Fourth, financial measures are contra-customers. Numerous companies have had massive layoffs in an attempt to cut costs; these layoffs significantly destroy the morale of the remaining employees. I heard an IBM employee say recently, "Now I'm expected to do three people's jobs. I just can't do it. I haven't even got enough time in the day to answer all my phone calls, so I just don't answer any of them." Unfortunately, cutting employee costs was often the least valuable cost to cut if operational efficiency was a goal; but profitability was the main goal, not operational efficiency. Employee costs are attacked first because they are the easiest to cut and have the most direct effect on the profit and loss statement, if you're looking for "short-term" financial performance.

Operational Measures of performance are more long-term oriented. And, as numerous publications make clear, operational performance achieves financial performance as a side-product.[11] The measure chosen should be the one most focused on the goals of the company; it should enhance these goals and move them forward. It is inappropriate to select all of these goals, or any group of them, if they are not focused on the ultimate objective of the company.

Quality is one of the fuzziest operational measures around. If you asked ten people to define quality, you'd get ten different definitions. Therefore, it is important to clearly define for your employees what quality is. Unfortunately, for most companies, quality has nothing to do with the customer, but instead is defined as "meeting and exceeding engineering specifications" or as "satisfying ISO 9000 requirements." Neither ISO, nor the engineers, have a very good understanding of what customers want. You can have an ISO certification and be producing useless junk, but as long as you are producing it to engineering specifications it is labeled as "quality." The only meaningful definition of quality is the one mentioned earlier: "producing a product or service so exciting to your customers that they wouldn't think about purchasing it from anyone else but you."

Productivity is the second fuzziest of operations measures, and it also requires clear definition if it is to be meaningful. The textbook definition of productivity is

$$Productivity = \frac{Output}{Input}$$

Output is net sales. However, input can be anything in the world, and often is. Productivity can be a measure of a fixed point in time, or it can measure change over time. For the United States, input is direct labor hours (or sometimes dollars). And the U.S. is by far the most labor-productive nation in the world. Japan focuses on total factor productivity or value-added productivity, both of which increase internal cash flow. As a result, Japan has enormous balance-of-trade advantages with most of its trading partners. Other countries, such as developing countries, focus on resource productivities, or productivity improvements, and show impressive numbers in these areas. So what good is productivity as a measure of performance? It is valuable if you select a productivity measure that focuses on your goal, and it is meaningful if you use productivity measures to compare yourself against yourself over time (referred to as *internal benchmarking*).

Concept Management introduces a new type of productivity measure — thinking productivity. The focus is on thinking smarter, not harder. An example of this is found in data collection. Having high productivity in data collection has no value if the data doesn't need to be collected at all. Thinking productivity will be discussed in later chapters in more detail.

Efficiency is an internal measure of performance to a standard. It is often used to evaluate labor performance and is used for incentive pay based on better-than-standard performance. Unfortunately, this measure rarely says anything about quality or material productivity, and tends to encourage employees to "pump out parts"; efficiency tends to become contra-"efficient." Quality and material efficiency should be part of this measure if it is to be of any value.

Throughput is a measure of quality units shipped to the customer.[12] This measure requires a coordinated effort among all areas: marketing, distribution, manufacturing, engineering, etc. This is a customer-satisfaction-oriented measure that can be very valuable, but it should not be used in conjunction with other measures such as efficiency, because they would conflict. With throughput, we should measure and motivate all employees with this same focus.

Inventory levels or inventory turns is critical in today's discrete manufacturing environment, where as much as 75% (and occasionally even more) of the value-added productive content is materials. Typical turns ratios in the U.S. are at about three to ten turns per year. Toyota turns its inventory for its automotive production facility every 3 hours. Over two shifts, that would be over 1,000 turns per year — a striking contrast to the United States' numbers. The carrying cost for the average American car in an organization that has about four turns per year is about $2,000.[13] This means that American vehicles have an automatic price increase of $2,000 for all vehicles, which pays solely for the interest costs of the inventory. We consumers are financing inventory inefficiencies through higher prices. But this should not be deemed as criticism; rather, it should be considered an excellent opportunity for someone with a competitive tendency to take on other American producers.

Value-adding requires a shift in thinking for the entire organization. Rather than viewing all transactions from a cost/profit perspective, we consider them from a benefit standpoint. For example, we are not searching for customer satisfaction, we are looking for opportunities to add to the customers' value, making the customers more successful in what they are trying to accomplish. We are looking for ways to help the customers beyond mere satisfaction; we focus in delighting the customer by making their interest our primary interest.

The counterpart to value-adding is waste elimination. In the customer case, we would never recommend anything for the customer that was a waste (non-value-adding) even if it would increase profitability.

Internally to the organization, we would look for opportunities to focus on value-adding. For example, unnecessary paperwork or unnecessary movement of materials is a waste and should be eliminated. This focus on value-adding permeates thought, so that if we are building a television set, and the tuner and power supply each cost $10 to produce, but the tuner sells on the open market for $15 and the power supply sells on the open market for $20, I might select to purchase the tuner, but I would always want to build the power supply myself to maximize my internal value-added.

Focusing on increased internal value-adding is one reason why the Japanese balance of trade is so strong. Low value-added items are produced off-shore, and high value-added content items are kept internal to the country and/or company.

Concept Management introduces a twist to the conventional perspective on value-adding. Concept Management adds a focus on value-added thinking — a focus on the elimination of wasted thinking. Conventional logic focuses on analytical thinking, which can often be a waste. Concept Management stresses that we should identify the purpose of the process that we are going through, even if it is a thinking or analysis process, before we proceed to waste a lot of time (wasted thinking). More will be discussed about this perspective on value-adding later.

Benchmarking is a comparative tool with which we check our performance against something else. Two types of benchmarking exist, external and internal benchmarking. External benchmarking is the comparison of your performance against your competitors'. This is valuable if you are trying to catch up, or are in a competitive environment. For example, the AT&T measure called *Customer Value-Added* is an external benchmark.

Some excellent tools exist for external benchmarking. For example, there are national tables that utilize the Standard Industrial Classification (SIC) codes. SIC codes are codes that classify every type of industry. If you search these tables for your type of industry and company size, you can compare yourself and your performance against the industry average in balance sheets, profit and loss statements, and financial ratios.[14]

Often, a more valuable form of benchmarking is internal benchmarking: you compare your performance against yourself over time. Your measures should focus on the goals of the organization so you can demonstrate improvement in these measures. Be your most challenging competitor!

Concept Management introduces a new twist to benchmarking. It introduces the "absolute benchmark," which is a goal-driven, goal-focused benchmark. Unlike the internal benchmark, which focuses on comparing you with your past performance, or the external benchmark, which attempts to equalize you with your competition, the absolute benchmark generates a long-term target for effective performance. The absolute benchmark will be discussed in more detail in later chapters.

Competitors — Obstruction to Change

Competition does one of two things: it either causes you to lose your drive toward change, or it causes bankruptcy, thereby forcing a major change. Competition causes us to lose our drive because we fear failure; in the back of our minds we worry that the changes we make may fail, making us less competitive and costing us customers. On the other hand, opting for no change may cost us our competitive edge. We need to realize that competitors are also required to change, and whoever comes out with the best strategic changes first will be the winner.

Strategic, competitor-focused changes are based on quality and time-to-market issues, also a form of quality if it is what the customer expects. Time-to-market effectiveness means that you manage your supply chain better than your competitors do. You are able to take new ideas, turn them into customer-usable products, and coordinate the efforts of the suppliers, distributors, and retailers, so that the new idea becomes available to the customer at a faster supply rate than your competitors can supply.

Future — Direction of Change

If you don't make dust, you eat dust!

The future is coming, whether we're ready for it or not, and it will change everything. We can either fight it by trying to maintain stability in our organizations (often referred to as organizational control), or we can anticipate it and take advantage of it through Concept Management. To plan for the future we need to

> Know where we want to be (the focus of this section).
> Develop a road map to get us to our goal.

In setting our goals (measures) we need to consider both their focus and format.

Focus

The focus of goals can take us in four different directions

1. Financial goals — short-term and conflicting
2. Operational goals — more long-term and effective
3. Employee-based goals — like FedEx, where treating the employees like customers defines the management style
4. Customer-based goals — where customer excitement about your product is the primary objective[15]

> *A lot of people climb the ladder of success only to find that its leaning against the wrong wall.*
>
> **Boyd K. Packer**

> *There are countless people who use the goal approach to climb the ladder of success — only to discover it's been leaning against the wrong wall*
>
> **S. Covey, A. Merrill, and R. Merrill,**
> ***First Things First*, Simon and Schuster, 1994.**

Focusing on just one goal (or goal set) is critical. Having multiple, and often conflicting, goals can be confusing and self-defeating. This is discussed in more detail in the chapters detailing the philosophy and procedures of Concept Management. Characteristics of good goals include

1. Participatively created by and matched to the employees
2. Shared with the employees
3. Nonconflicting
4. Allow for and encourages change
5. Simple (easy to understand and remember) but not simplistic
6. Precise
7. Measurable
8. Uncompromised
9. Focused
10. Achievable yet challenging

Format

In the development of a goal there are several stages:

Defining the core competencies — What is it that you are good at, better than anyone else? (one sentence)

The vision — Goals 30+ years into the future that magnify and don't ignore the core competency (one sentence)

The mission — Goals 10+ years into the future with specific dates and milestones attached, focused on the vision (a couple of sentences)

The strategy — 5 to 10 years of annual plans working us step by step toward the mission

The plan of operation — The current year of the strategy detailed out with precise budgets and forecasts.

The strategy should be developed for at least two levels: the corporate level (focused on the corporate mission statement), and the business unit level (focused on the corporate strategy). At each level we detail a plan that discusses procedure in each area (if relevant) as we move toward our mission:

People
   First Employees,
      Education and training
      Empowerment
      Teamwork
      Organizational structure, staff functions
   Then Customers,
      Involvement
   Then Vendors,
      Integration
   Then all other stakeholders
Integration
   Elimination of horizontal and vertical barriers
   Information
Globalization
   Internationalize but localize
Measurement
   Internal Performance
      Quality, productivity, efficiency
   External Performance
      Adding value to society
      Customer-perceived quality
      Market share
   Internal Factors
      Capacity
      Equipment
      Operational performance

    External Factors
        Competition
        Economic conditions
        Government regulation
    Continuous Change Process Focused on Adding Value
        Elimination of Waste
        Identifying Strength and Weaknesses
        Identifying Opportunities and Threats
    Time-Based Competition
        Time-to-Market Strategy
        Supply Chain Management
    Technology
        Funding
        Facilities and Equipment
        Product and Process Innovations

Building goals is often taken too lightly. Without working through a goal development process, a meaningful measurement/motivation process cannot exist; employees simply move along without making progress.

Failing to Plan is Planning to Fail

There is much valuable literature on how to develop goals. One excellent source is the book *World Class Manager* by Gerhard Plenert. This book discusses in detail both format and focus.

Integration — Success Behind the Change Process

Through integration everyone and everything works together. Managers are not merely bosses, they are leaders and facilitators by example. They work side by side with the employee. Managers are, as with FedEx, evaluated by the employees and can even be fired by the employees. Managers are not possessive of power and, as in Semco of Brazil, are willing to admit that perhaps the employees know more about the operation of the facility.

Integration means that barriers between departments no longer exist. If you have a problem or a question or are just interested in something, you find out about it without having to go through a chain of command.

Accountants can sit down with engineers and watch what they do. Inventory control clerks can check with a drill-press operator and find out how badly he or she really needs the parts.

Integration also focuses on time-to-market efficiencies. The objective is to get the new product idea to the customer faster. The closer all departments work together without bureaucracy and paperwork, the more efficiently you can get the product to the market.

Integration also means supply chain management. This coordinates the efforts of everyone involved, from the vendor to the distributor, through the facility, and out to the customer, thus eliminating waste in steps. Several years ago the Building Industry Association of San Diego County ran an experiment on how quickly a three bedroom, two bath house could be built using standard materials. Several builders participated, but the winning house was built in 2 hours and 45 minutes. This rapid construction project required 700 people divided into teams. The teams rehearsed and prepared, and when the competition began, they did the rough plumbing in 8 minutes, and set the roof in just over 9 minutes.[16] Is it reasonable to have 700 people work on one project? Probably not, but it is also unreasonable that the average house of this type would normally take 165 days to construct. If customer responsiveness is important to you, then supply chain management is a critical element.

Integration also focuses on information. Information can be advantageous, as in Wal-Mart's integrated logistics Electronic Data Interchange (EDI) system, which informs Wal-Mart's distribution centers of the previous day's sales while the distribution centers are already loading the trucks for the next day. Information can also be a waste, as in the case of Ricardo Semler, who threw all the information systems out of Semco, feeling they were causing an unnecessary degree of bureaucracy; and as a result, he significantly cut operating costs and reduced lead times. The key is to collect data and generate information only when it is value-adding.

Non-trust systems (systems installed to prevent mistakes or to prevent fraud) are nearly always a waste, yet most of our systems are non-trust based. KAO of Japan reduced their accounting staff from 150 employees down to 7 employees by eliminating non-value-adding non-trust systems throughout their organization. They developed a system called the regularity concept to accomplish this feat (a further discussion of this concept occurs in Chapter 6). Additionally, data collection for the sake of data collection, because all data is believed to be good, or because some day this data may be valuable, is also a waste. Don't run your business on the basis of old data, base it on future projections. Data and information should be value-adding by helping you achieve your goal. Information systems should improve productivity and quality directly; all else is a waste. You need to check your information systems and identify their real purpose. Perhaps they were installed at a point when they were valuable, but now they simply exist because of tradition. In that case, get rid of then.

Concluding World Class Thoughts

It is enormously difficult to do justice to a concept such as World Class Management in one chapter. I have nevertheless attempted to give you a feel for what World Class Management entails. For more details, review the referenced materials. In conclusion, let me leave you with a few thoughts on customer satisfaction:

1. Anyone, from the CEO to the individual who sweeps the floor, who does not directly help a customer or who does not directly help someone who directly supports a customer is a waste. Multiple levels of corporate hierarchy are wasteful and not value-adding, unless, of course, your corporate goal is to keep as many people employed as possible.
2. Never fire anyone; it destroys trust. Always try to find a new position for employees somewhere else in the company; but, if that is not possible, help your terminated employees be employable by training them and helping them find new jobs. The trust and respect this earns for you will be invaluable.
3. Ask yourself why the employees who are the least trained, least paid, least experienced, least motivated, have the least understanding of the company and often care the least about the company's success are also the employees who have the most direct contact with and influence over your customers? Whether we are talking about a bank clerk, a grocery store teller, or a front office secretary, these are the individuals whose attitudes will make a customer want to come back or go away. Here I stress two things:
    a. Get managers directly involved with customers.
    b. Motivate and train anyone involved with the customer (this should be everyone) to make sure they make a good impression with their knowledge about the company, so that they portray an enthusiasm that shows they care about the customer.

At this point I have filled you with a multitude of ideas about becoming World Class. The ideas in this brief summary will be integrated, along with the ideas about Breakthrough Thinking (Chapter 2) and the ideas about Change Models (Chapter 4), to help us develop the integrated principles of Concept Management (Part II).

> *"Always keep in mind that your own resolution to success is more important than any one other thing."*
>
> **Abraham Lincoln**

Endnotes

1. Gerhard Plenert, one of the authors, and his family spent one year teaching and doing research in Kuala Lumpur, Malaysia for Universiti Malaya.
2. Taken from the FedEx brochure *Quality Profile*.
3. The company name has been withheld as a courtesy to those involved.
4. Some excellent material exists on teaming and team building. Let me recommend a few: Pope, Sara, *Team Sponsor Workbook*, Cornelius and Associates, Columbia, SC, 1994; Pope, Sara, *Team Leader Workbook*, Cornelius and Associates, Columbia, SC, 1996; Mears, Peter and Frank Voehl, *Team Building: A Structured Learning Approach*, St. Lucie Press, Delray Beach, FL, 1994.
5. Taken from the FedEx brochure *Quality Profile*, p. 3.
6. Senler, Ricardo, *Maverick*, Warner Books, New York, 1997, p. 150.
7. Baker, Stephen and Larry Armstrong, "The New Factory Worker," *Business Week*, September 30, 1996, p. 59–68.
8. Covey, Stephen R., *The Seven Habits of Highly Effective People*, Simon and Schuster, New York, NY, 1989, p. 58.
9. Anonymous, "Happy Labor Day," *Time*, September 4, 1995, p. 21.
10. This information was taken from a presentation by Lyle Tippetts in Guatemala, 1994. Lyle is the Quality Director of the Caribbean/Latin America region of AT&T.
11. The following books discuss the value of operation goals in more detail: Goldratt, Eliyahu M. and Jeff Cox, *The Goal*, North River Press, Croton-on-Hudson, NY, 1986; Goldratt, Eliyahu M., *The Haystack Syndrome*, North River Press, Croton-on-Hudson, NY, 1990; Goldratt, Eliyahu M. and Robert E. Fox, *The Race*, North River Press, Croton-on-Hudson, NY, 1986. These books compare financial and operational goals in more detail: Plenert, Gerhard, *The Plant Operations Handbook*, Business One Irwin, Homewood, IL, 1993; Plenert, Gerhard, *World Class Manager*, Prima Publishing, Rocklin, CA, 1995.
12. See earlier references by Eli Goldratt for more details.

13. If you take a $30,000 vehicle, at four turns per year, the vehicle will be in inventory ¼ of the year. Additionally, there is another ¼ year at the vendors that supply the components, and still another average ¼ year at the dealership. Then the $30,000 vehicle has ¾ of a year's inventory assigned to it. If we take this times the interest rate of 10% (estimate), we get

$$\text{Carrying Cost} = \text{Time} \times \text{Interest} \times \text{Cost}$$
$$\$2{,}250 = 0.75 \times 0.10 \times \$30{,}000$$

    This number is underestimated because all the delays in the supply chain are not considered. For example, the shipment of parts from overseas can take weeks or even months. Also, the interest rate should be replaced by the opportunity cost rate (the rate we could be earning on the money if we invested it) rather than the bank financing cost. A more appropriate opportunity cost rate would be 20%, which would double these cost estimates, or more appropriately, "lost profitability."
14. One of the most popular tables of industrial ratios is the *Industrial Norms and Key Business Ratios* put out by Dun and Bradstreet Information Services. This report will give all the average ratios, the average profit and loss statement, and the average balance sheet for 1 year for each SIC code. It is available in most of the business libraries.
15. A detailed discussion of each of these goals and how they work can be found in Plenert, Gerhard, *The Plant Operations Handbook*, Business One Irwin, Homewood, IL, 1993, Chapter 6; Plenert, Gerhard, *World Class Manager*, Prima Publishing, Rocklin, CA, 1995, Chapter 3.
16. Anonymous, "Can't You Hammer any Faster?" *Fast Company*, August/September, p. 40.

Chapter 4

Change Methodologies

That's perfect! We've lost our way but we're making good time.
From the movie *City Slickers*

I found it quite interesting to watch the launch of a space shuttle. At first there was a lot of loud noise; the earth shook; there was flame, flashes, smoke, and bright lights from the fire; but nothing moved. Then, slowly, the shuttle's engines started to lift it off the ground, initially by just inches, then more and more, until the engines lifted the shuttle skyward to new heights. Isn't that a lot like change? Whenever change is introduced you first get the rumble; the earth shakes; there's a lot of flashes and noise. Then, slowly, as individuals start to see the benefit of the changes, success brings us to new heights. Wouldn't it be nice if we didn't have to go through the trauma caused by newly introduced change? That's what change models are for: they help us systematize and standardize the change process so that it doesn't cause quite as many earthquakes.

The focus of leading-edge organizations has shifted from one of seeking stability to one of managing change: change in products, components, demand, resources and their availability, changes in operational technology, competitive product makeup, competition, etc. Continuous improvement (change) should include[1]

Product innovation
Process innovation
Technology innovation
Time-to-market innovation
Marketing innovation
Etc.

Uncontrolled and undirected change can be even more disastrous than no change. What we need is to be able to manage and at the same time stay ahead of the change process. We need to change ourselves faster than external forces push us to change, to focus on a target while at the same time maintaining corporate integrity.

To manage change we need to incorporate change models that facilitate the change process. Most change models contain some label of quality in them. Quality improvement programs often focus on continuous change, but "quality" doesn't fully define everything wanted by the change process. Nevertheless, terms like Total Quality Management (TQM) and Quality Functional Deployment (QFD) are change processes, and "quality" is a characteristic of how they are used, not how they were designed. In reality, like all change models, they can be focused on all positive, goal-direct changes in all areas of the organization including quality, productivity, efficiency, financial improvements, etc.

Before we discuss some of the change models specifically, let's first discuss the psychology behind change. There are two sequences for you to complete. These should be easy to do *if* you look beyond traditional methods

Change Methodologies

of analysis. To solve these you need to be able to expand your mind beyond the tendency to make everything complicated. You need to look at the "big picture" to find the easy and the obvious, and you need to expand your experience base. The sequences are (for those of us that need a little help, the answers are in the Appendix at the end of this chapter)

$$O\ T\ T\ F\ F\ \_\ \_\ \_$$

$$J\ F\ M\ A\ M\ \_\ \_\ \_$$

Are we looking for changes in areas where it's convenient, or are we looking in the areas where we'll get the most benefit for our efforts? Often we take the easy way out when it comes to confronting change — avoidance. But why do we avoid change? The avoidance of change can be summarized in one word: **Fear.** Resistance to change should not be considered irrational, especially if the change directly affects our job function. Don't fear resistance, work with it. Not all change is good change; sometimes the way change is instigated makes the change bad. And at other times, the change fails, no matter how hard we try. This returns to the discussion of ownership in the last chapter. If employees have ownership in the change process, no matter how ill-conceived the change is, they will make it work. Conversely, if employees do not have ownership in the change, no matter how excellent and perfect it is, it's doomed to failure. But without change we are sure to fail. We must manage our way around the resistance caused by the fear of change.

In production processes, the Japanese use rocks in a river to symbolize resistance to change. Water flows smoothly down the river until it encounters the rocks. The water must work its way around the rocks in order to successfully move down the river. In implementing change, the rocks we encounter most often do not come from the employees, but from management. Consultants explain the source of the greatest resistance when they say, and quite rightly, that "the hardest rocks wear ties (or nylons)."

The toughest resistance to change comes from managers committed to the traditional way of doing things. They learned to do it that way in school, or they've always done it that way, and they don't understand why they need to change now. The line workers are used to being jerked around by new-fangled ideas. They have become used to the changes from management but they are not necessarily excited about or motivated to work with them. Often, like the frog in the picture, one individual's resistance to change holds other individuals back and keeps not only them, but the entire organization, from making the jump.

Resistance to change should be anticipated and worked with by helping those that fight the change to understand the change and the reason for it, and to develop "ownership" of the change and the change process. If employees feel ownership in the change, resistance will greatly decrease; ownership is developed through the understanding gained by education and training.

Change can be better understood by considering the change model. In this model we start from our current state (Stage A) and we see that change, like a space shuttle launch, initially causes a drop-off of results (Point X to Stage B). This could be because of resistance due to a lack of understanding, or it could be because of a learning curve. Then, as the change is implemented and given time to settle in (Stage C) we see the recovery of the losses that resulted from the initial change implementation (Point Y). From here on, the implemented change is making life better for us (Stage D) until the change slowly works us to a higher steady-state plane of achievement (Stage E).

Change and the change management process also require commitment, much like the bacon and eggs you ate for breakfast. In that breakfast the chicken was involved, but the pig was committed. The source of any change initiative — and it's usually management — must be committed to the change process. Commitment sets a desire for change throughout the organization. Next we need goals (Chapter 3) in order to give the change process direction; then we need training to know what to change, how to change it, and when to change it. Understanding change begins with understanding the change process.

Change Methodologies 55

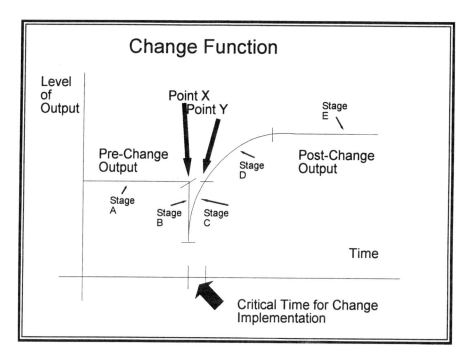

Change models supply us with a methodology for implementing changes. They offer us a process for organizing the change so that its implementation can be understood by everyone involved. In this chapter we will discuss the two most popular, leading-edge change models, and compare their advantages and disadvantages. Then we will consider a variety of usable tools in the following change models to facilitate positive, growth-oriented change:[2]

Total Quality Management (TQM)
Process Reengineering (PR)

In Appendix 4B at the end of this chapter, we have also considered the following change implementation tools:

Quality Functional Deployment (QFD)
ISO 9000
The award processes
KAIZEN
Just-in-time (JIT)
Process mapping (flow charting)
Quality circles

Statistical process control (SPC)
Quality gurus
Manufacturing resources planning (MRP II)
Enterprise resources planning (ERP)
Time-to-market and cycle time management
Supply chain management

These tools work within the structure of TQM and may or may not be applicable for your particular environment. They are mentioned in the appendix because you may wish to consider them as part of your specific change model. Extensive detail about these tools can be found in the literature.

After a discussion of the change models, we will then discuss how Concept Management and Concept Creation are both needed for a change model to improve the implementation process.

Circumstances — what are circumstances?
I make circumstances!

Napoleon Bonaparte

Total Quality Management (TQM)

TQM is the tool that should be used to formalize the search for and the implementation of change. TQM has gone through various translations and currently exists in many versions. We need to more clearly redefine TQM as a change process. In this book we will attempt to develop a generic TQM breaking free from the traditional TQC (Total Quality Control) limitations. Traditional TQC systems do not focus on change processes. Rather, the TQC focus is on maintaining standards and is based on a control paradigm. Unfortunately, this limiting control paradigm has been incorrectly carried into TQM as part of TQM philosophy. We need to use TQM as the change implementation model that it was intended to be all along.

In the U.S., TQM has fallen into disfavor because of its analytical approach to change. The analysis process is deemed too slow to be competitive. However, in light of the innovation in the analysis process that Breakthrough Thinking offers, we need to revisit our use of TQM. Perhaps it is the best change management tool after all.

The are two major aspects to TQM: a philosophical aspect, and an operational aspect. From the philosophical we get guidelines and from the operational we get techniques. We will discuss both in detail. Later we will attempt to integrate TQM into Concept Management as our change management organizer.[3]

The Philosophical Elements of TQM

In the words of Karen Bemkowski "Simply put, TQM is a management approach to long-term customer satisfaction. TQM is based on the participation of all members of an organization in improving the processes, products, services and the culture they work in."[4] Traditionally, the philosophy of TQM could be stated as, "Make sure that you're doing the right things *before* you worry about doing things right!"

Total Quality Management (TQM) focuses on careful, thoughtful analysis. However, the analysis should be creative, innovative, and "innoveering" oriented. We want to make sure that we are implementing positive, goal-focused changes before we move a muscle.

TQM is an enterprise-wide change model. Some people define TQM as simply making the "entire organization responsible for product or service quality." This is the way TQM is defined in many organizations and it encompasses everything and anything during the change process.

To some, TQM is a behavior-based philosophy of motivation and measurement. TQM does, in fact, require a cultural shift for all members of an organization in that it uses an entire philosophy about how entire businesses should be run. TQM is filled with ideas and attitudes:

- Attitude of desiring and searching out change
- Think culture — move from copying to innovating
- Focus on the goal
- Measurement/motivation planning
- Top to bottom corporate strategy
- Companywide involvement
- Clear definition and implementation of quality
- Education, training, and cross-training
- Integration and coordination
- Small, step-by-step improvements

In TQM we become excited about changes. We look for opportunity to change because change means that we are becoming better. To be a TQM organization is to become an organization that wants to be the best, and realizes there is always room for improvement.

Success stories for TQM can be found in settings all over the world, as measured by the successful implementation of change. This change can be in the form of new technology or in the correction and improvement of old technology. Often, a successful TQM project results in the ability of employees to work more effectively together. The result is that the measurement of TQM success tends to be an internal success story, and not always externally comparable.

Success in TQM can be found in large organizations such as PETRONAS, the national petroleum corporation of Malaysia, where, because of its successes, TQM implementation is moving forward on a companywide basis. Through its systematic implementation of changes, TQM was instrumental in helping to win the Deming Award, a Japanese Quality Award, for Florida Power and Light, a U.S. producer of electricity. TQM is receiving attention in Latin America, Europe, and Africa through national drives for quality, like Central America's national congresses on Total Quality held in Panama and Guatemala. The AT&T program that won numerous Baldrige Awards is one of the success stories of individual companies.

There is also a specific, proceduralistic version of the definition of TQM. Perhaps the best way to understand TQM is to look at this process and its significance.

The Procedural Elements of TQM

In operationalizing TQM there are several elements of structural importance:

- The TQM coordinating team (Quality Council)
- The three "P" teams — cross-functional teams
- The TQM project implementation steps
- Training programs
- Measurement and feedback
- Showcasing
- Team building
- Systematic Problem Solving (SPS)

TQM implementation starts with a coordinating team, often referred to as a "Quality Council." This is a team composed of high-level corporate leaders from all the functional areas, usually at the vice-president level. This team is appointed by the CEO and operates under his/her direction. The CEO actively directs the endeavors of the team, and is often an active team member. This Quality Council is then responsible for organizing, chartering, and measuring the performance of the other TQM teams within the organization. It oversees the installation, training, performance, and measurement of the other teams. This team aims to keep all teams focused on the corporate goal and vision.

The Quality Council will organize three different types of teams referred to as the cross-functional "three 'P' Teams": "p"rocess, "p"roduct, and "p"roject teams. The process teams are ongoing, continuous-improvement teams organized at various levels of the organization. They look for improvements in the organization's functioning processes. There would be teams organized to analyze production processes as well as information processes, such as the accounts receivable system. These teams should be composed of both "insiders" (individuals familiar with what is under study) and "outsiders" (individuals unfamiliar with what is being studied). The insiders know and understand existing functions and operations. The outsiders challenge the status quo.

The second of the three "P" teams are the product teams. These teams are cross-functional but focus on a specific product, product line, or service. They are customer and vendor interface teams specifically oriented toward the development of new products and the improvement of existing products. Their life span is the same as the life span of the product they represent.

The third of the three "P" teams are the project teams. These teams are limited-life teams set up to focus on a specific project, such as the construction of a new plant, or a computer installation. These teams may be the result of a specific process or product being targeted, or they may be set up to research something the general management team is interested in developing or improving.

Once a team has been organized and its team members selected, the team is given a charter by the Quality Council. This charter defines

- The goals and objectives of the team
- How the team is measured
- What the team is empowered to do
- How the team will participate in the benefits of the improvements (gainsharing)

With the team organized and focused, it is now ready to begin operation. The members are trained in the techniques that go along with teaming, which include the analysis tools that should be used, and the standardized procedural steps that should be followed.[5] These include the TQM project implementation steps, which are as follows:

1. Identify problems (opportunities)
2. Prioritize these problems
3. Select the biggest bang-for-the-buck project
4. Develop an implementation plan
5. Use operations research and MIS tools where appropriate
6. Develop guide posts and an appropriate measurement system
7. Train
8. Implement
9. Feedback — Monitoring — Control — Change
10. After successful project implementation and ongoing status, repeat cycle

The first function of the team is to identify its own function within the stated charter. If you are one of the three "P" teams, your team's charter is laid out for you by the Quality Council. If you are the Quality Council, this charter is laid out for you by the CEO and is aimed at the focused goals of the organization. After understanding the charter, the team will then search for problems that prevent the organization from achieving this charter. A better wording for the negative "problems" would be to say that we search for "opportunities for improvement." We are not just trying to correct negative effects, we are looking for techniques or tools that will allow us to become better and possibly even the best.

Next we take these problems (opportunities) and prioritize them on the basis of their effect on the team charter, which should be focused on the goals of the organization. We do a type of ABC analysis (80–20 Rule, or Parieto Principle) to determine which change would have the greatest effect. Then we select the biggest bang-for-the-buck project and develop an implementation plan. This implementation plan will contain guideposts based on an appropriate measurement system that points the team toward achieving its charter.

Training of the implementers and users is critical, or else the planned project is doomed to failure. This training makes future users comfortable with the changes. It also offers a bit of ownership, since the planned users will now feel comfortable with the changes.

The next step is implementation, and it should be a trivial process if all the planning and training steps are performed carefully. Part of the implementation is the installation of feedback, monitoring, and control mechanisms, as laid out in the implementation plan. Careful monitoring allows for corrective changes to occur whenever necessary.

After successful project implementation, and seeing that the ongoing status of the project is functioning correctly, the team repeats the implementation cycle, looking for more opportunities for change. If this process is performed correctly, the list of change opportunities should become longer with each iterative cycle. This means that your team is now open for newer and broader opportunities for change.

A few elements of a TQM project that seem to make a difference between success and failure include

- Training programs
- Appropriate measurement and feedback mechanisms
- Showcasing (a "sure thing" TQM implementation)
- Team synergy development

The Good News About TQM

TQM was the first stage of realizing that we need to take "quality" (or the search for positive change) out of the quality department and make it a companywide program. TQM is a strategy toward continuous, corporatewide change; it is a philosophy; it is an operationalized process; it is a fad. It becomes a fad if we expect quick results and become disenchanted because we are not "like the Japanese" in the first 2 months. TQM is a strategy for becoming leading edge and World Class.

TQM differs from the other quality tools such as Total Quality Control (TQC), Statistical Process Control (SPC), or In-Line Quality Control (ILQC) in that it is not as directly focused as these other systems on a specific procedure. Rather, TQM is a continuous search for problems (opportunities) to eliminate waste and add value in all aspects of the organization, and it makes these improvements one small step at a time.

Despite its slowness, TQM has been extremely successful internationally and is getting ever-increasing attention. References to TQM and its leadership abound.[6] TQM is a very specific process improvement step in a drive toward World Class status.

Some Shortcomings Of TQM

One of the biggest downfalls of a TQM system, as far as the U.S. is concerned, is how long it takes to implement the change. Concept Management, through its use of the Breakthrough Thinking paradigm, dramatically reduces the lead time for change implementation. This was discussed in Chapter 2 and will be integrated into Concept Management as part of Chapter 5.

Often in the U.S., a decision is made to implement a change, and then we start worrying about how to implement the change. *Systematic Problem Solving* (SPS) is a procedurization of the change process that has become an integral part of TQM. There is no one perfect model for how this change procedurization should be set up, but there are a few good examples:

1. Florida Power and Light when they won the Deming Award
2. The AT&T SPS process
3. The T-Model.

Systematic Problem Solving (SPS) at Florida Power and Light

In the Florida Power and Light (FP&L) case, the SPS process it used when it won the Deming Award was referred to as its "quality improvement story," a series of standardized steps that are used to organize and document the change process. The steps are

1. Team Information — Here they develop a Team Project Planning Worksheet.
2. Reasons for Improvement — This is a graphic and flow-charted look at why an improvement is desirable (purpose analysis in Breakthrough Thinking)
3. Current Situation — Where we are now.
4. Analysis — A search for root causes in the FP&L case.
5. Countermeasures — Here FP&L will develop a Countermeasures Matrix and an Action Plan.
6. Results — FP&L will use Parieto diagrams and Before and After graphs to validate that the changes are occurring.
7. Standardization — This section establishes a documented procedure for the ongoing operation of the change.
8. Future Plans — This is a review of what was learned by this change process. It follows a philosophy of Plan–Do–Check–Act.

Systematic Problem Solving (SPS) at AT&T

AT&T uses a methodology that includes tasks performed in four distinct stages, which are

1. Ownership — Team responsibility for the activities.
2. Assessment — Clear definition of the process.
3. Opportunity Selection — Analyze how process problems affect customer satisfaction and rank them in order of opportunity for improvement.
4. Improvement — Implement and sustain the change.

Change Methodologies 63

The Ownership, Assessment, and Opportunity Selection stages are considered management processes. Then, based on the overall four stages, grouped under management and improvement, AT&T developed a series of steps called the management and improvement steps, which focus on the SPS process:

1. Establish process management responsibilities
2. Define process and identify customer requirements
3. Define and establish measures
4. Assess conformity to customer requirements
5. Investigate process to identify improvement opportunities
6. Rank improvement opportunities and set objectives
7. Improve process quality

Note that the AT&T process follows the Japanese model closely. More detailed information is available about this process in publications by AT&T.[7]

The T-Model

The third systematic problem solving model discussed is called the T-Model, a systemized model for change that has a philosophical as well as a procedural aspect. It has its roots from Breakthrough Thinking and adapts easily to it. Philosophically, this model looks for rapid, continuous improvement — change implementation. Procedurally, the T-Model follows a series of basic rules, or steps:

1. Define area of change
2. Define the purpose of the change — don't ask why the change should be made and don't do an analysis of the change and its requirements. The purpose of the change needs to be defined first.
3. Evaluate the purpose — does it eliminate waste and improve the value-added component of the product?
4. Define the constraints — environmental, customer, cultural, etc.
5. Evaluate the techniques available for solving the problem
6. Implement the change
7. Monitor, get feedback, and collect data
8. Engage in corrective action — here we react to the feedback. If the feedback is not what we want, then we return to step 2 above and rethink the corrective action.

The T-Model is more general than the Florida Power and Light example, because it is not applied to a specific situation — no specific tools are assigned to each of the steps. However, as the T-Model is applied to a specific example, such as in the Florida Power and Light situation, it would become more focused, detailing specific tools and procedures that should be used.[8]

TQM Summary

TQM is an excellent change implementation tool. By adapting Breakthrough Thinking principles to the analysis process, and by broadening the scope of TQM to incorporate all change areas, TQM becomes a perfect tool for continuous improvement implementation as required by Concept Management. By integrating TQM principles with Breakthrough Thinking Principles, we eliminate the extensive data collection that traditional TQM models have required in their quest for "root cause analysis" solutions. TQM becomes fast and extremely effective in the development of a World Class environment.

> *We are not creatures of circumstance,*
> *we are creators of circumstance.*
>
> **Benjamin Disraeli**

Process Reengineering

Process Reengineering (PR) is rapid, radical change that originated in the United States. It is work elimination and *not* downsizing, which many companies are using it for. It is positive, growth-focused change, looking for opportunities to eliminate waste and improve value-added, often through the implementation of technology.

In 1994, $32 billion was invested in reengineering, but two-thirds of the reengineering projects will fail. Why? Because the change process builds up resistance, thereby forcing its failure. Failure also happens because PR is used as an excuse for downsizing, often resulting in the elimination of critical employees that will be difficult to replace. The downsizing process was not carefully thought through, and this rush renders disastrous results.[9] Employees feel resentment rather than ownership in the change process. PR is dictated from the top and tends to be fear-based.

Change Methodologies

However, just like any tool, some extremely positive aspects to process reengineering make it worthy of our attention. The first is that PR focuses on change implementation at the top of the corporate hierarchy; it generates more of a top-down, change-committed culture. It focuses on process-oriented change, which is a Japanese trademark.

PR's focus on the process emphasizes that the process, not the product, holds the secrets for the most dramatic improvements within an organization. PR focuses on an "all-or-nothing proposition that produces impressive results." PR is defined as "the fundamental rethinking and radical redesign of business processes to achieve dramatic improvements in critical, contemporary measures of performance, such as cost, quality, service, and speed."[10]

The principles of reengineering include the following:

- Organize around outcomes, not tasks.
- Have those who use the output of the process perform the process.
- Subsume information processing work into the real work that produces information.
- Treat geographically dispersed resources as though they were centralized.
- Link parallel activities instead of integrating their results.
- Put the decision point where the work is performed, and build control into the process.
- Capture information once, and at the source.

The three R's of re-engineering are

- Rethink — Is what you're doing focused on the customer?
- Redesign — What are you doing? Should you be doing it at all? Redesign how it can be done.
- Retool — Reevaluate the use of advanced technologies.

Like TQM, the focus of the reengineering effort is the team. Departments are replaced by empowered process teams, executives change their roles from scorekeepers to leaders, organizational structures become flatter, and managers change from supervisors to coaches.[11]

Although PR has many of the procedural characteristics of TQM, it is more philosophical than TQM. PR focuses on being competitive via the rapid and the radical, stressing the process as the key to successful change. Numerous books are available that discuss the philosophy of PR, but the best is still the original, by the gurus of Process Reengineering, Hammer and Champy. SME/CASA put out an excellent booklet focusing on manufacturing processes that can be reengineered. *OR/MS Today* also has an excellent article that discusses first- and second-generation reengineering programs.[12]

Process Reengineering was introduced into Japan and thrown out within 1 year. The "rapid and radical" was considered to be too unplanned and unorganized, making TQM still the most strongly favored change implementation methodology in Japan in spite of shortcomings in its analytical approach. Process Reengineering and its management approach is considered to be similar to trying to squeeze water out of a dry towel, the idea being that, once upon a time the towel was wet, and squeezing it produced a lot of water. However, after so much squeezing, the towel has become dry. Squeezing this same towel may generate a drip; however, we are surrounded by piles of wet towels and we don't even see them; we are so obsessed with squeezing the towel we are holding. This towel that we are obsessed with squeezing (at least in the U.S.) is the labor towel; the towels we are ignoring are inventory, or maintance, or facilities resources, or machine efficiencies, etc. It is time to look at the other towels around us and not attack the labor resource through programs such as downsizing. Concept Management helps us to refocus on new towels.

Comparison of the Change Models

There is no "best" change model. However, for Concept Management to work most effectively, we believe that Total Quality Management (TQM) offers the most flexibility. With TQM we have a structured approach to change implementation; however, we can pull out the traditional, scientific approach toward analysis and implement Breakthrough Thinking (BT) as the creativity engine. With BT and its purpose expansion, we eliminate the slowness of the root cause analysis that exists in traditional TQM. As a result, the traditional criticism of TQM is eliminated and the change process is significantly faster and more results-oriented.

Process Reengineering is too authoritarian in its management and implementation approach to be World Class. It focuses on *doing* before *deciding* and doesn't involve the participation of everyone, as do TQM and World Class Management (WCM) styles. A World Class change model should focus on effective, customer- and employee-oriented change management that offers competitive innoveering (the name for the innovation engineering organization established by Walt Disney Corporation) strategies. World Class change management is Total Quality Management. TQM offers the most structure and tends to be the least resistive.

Change Methodologies

At this point we are committed to the following:
Concept Focusing with World Class Management
Concept Creation using Breakthrough Thinking
Concept Engineering with BT and TQM
Concept In, which is the marketing of the new concept
Concept Management with BT, TQM, and WCM, which integrates everything from Concept Focusing on up into one change opportunity system

The development and integration of Concept Management will begin in the next chapter.

Summary

Change is a continuous, ongoing, never-ending challenge of which we must eagerly take control. In the case of the mountain climber who slipped, he is grateful for the one spike that held, not for the 99 that "almost held." Similarly, we are grateful for the changes that successfully make a difference, not for the faddish changes that only complicate our lives. Therefore, a TQM change-planning process helps keep us focused and integrated, and becomes a critical tool in the development of Concept Management.

The Harder I Work, the More I Live.

George Bernard Shaw

Appendix 4A

The solution to the two sequences are

O T T F F *S S E*
One, Two, Three, Four, Five, *S*ix, *S*even, *E*ight

J F M A M *J J A*
January, February, March, April, May, *J*une, *J*uly, *A*ugust

Appendix 4B — Change Implementation Tools

The following are tools that can be used within the TQM framework in order to develop effective change. This discussion is intended only as an introduction to these concepts. More detailed discussions are available through a variety of sources.

Quality Functional Deployment (QFD)

QFD is the implementation of a continuous improvement process that focuses on the customer. Developed at Mitsubishi's Kobe Shipyards, it focuses on directing the efforts of all functional areas on a common goal. In Mitsubishi's case the goal was "satisfying the needs of the customer." Several changes were instituted in order to accomplish this, such as increased horizontal communication within the company. One of the most immediate results was a reduced time-to-market leadtime for products.

QFD systematizes the product's attributes in a matrix diagram called a *house of quality,* and highlights which attributes are the most important to a customer. This helps the teams throughout the organization to focus on each goal (customer satisfaction) whenever they are making change decisions, like product development and process improvement decisions.

QFD focuses on

1. The customer
2. Systematizing the customer satisfaction process by developing a matrix
3. Empowered teaming
4. Extensive front-end analysis, which involves 14 steps in defining the *house of quality*

QFD has been widely recognized as an effective tool for focusing the product and the process on customer satisfaction, and much has been written on the subject.[13]

ISO 9000

ISO 9000 is a certification process often advertised as a model for change and improvements; however, the ISO 9000 process tends to focus on stability. The ISO standard was developed in Europe for the European integration in an attempt to standardize the quality of goods coming into Europe. For many companies it seemed like a trade barrier that attempted to keep companies out of Europe. The reason is that ISO 9000 focuses on quality in the internal process of the organization, assuring that what was designed is what is actually built. It does not focus on the customer. Nevertheless, the ISO standard has become an international standard for quality and systems performance that many companies are using.

ISO has come to define quality, not change. It is a set of standards for quality based on two main foundations:

1. Management responsibility and commitment to quality which should be expressed in a formal policy statement and implemented through appropriate measures
2. A set of requirements that deal with each aspect of the company activity and organization that affects quality[14]

ISO can be used as a standard for improvement, and the ISO quality system requirements can become the focus of change systems. In this way, ISO criteria can be integrated into a change process.

Award Processes

Award programs, such as the Baldrige, Deming, and Shingo Prizes, have an excellent base of standards from which to build change processes. Like the ISO criteria, these award program criteria are an excellent basis for developing a focus for your change program. For example, the Shingo Prize organization focuses on continuous improvement processes through total quality systems. The Deming Prize focuses on demonstrated improvements resulting from a continuous improvement process. The Baldrige Award has a list of improvement criteria for award evaluation:

1. Leadership
2. Information and analysis
3. Strategic quality planning
4. Human resource utilization
5. Quality assurance of products and services
6. Quality results
7. Customer satisfaction

Within these seven categories there are 33 examination items and 133 subitems. As with the ISO process, the award process is not a change process, but it greatly assists an organization in establishing the criteria that should be incorporated into a effective change model. Going through the award process motivates the development of effective change procedures.[15]

Kaizen

The Japanese model for the continuous change process is called *kaizen*. It suggests that every process can and should be continually evaluated and continually improved. The primary focus of the improvements is on waste elimination, including

> Process time reductions
> Reduction of the amount of resources used
> Improved product quality

Kaizen is a guide to focused, continual improvement. Most Western quality textbooks mention it, but the concept is so fuzzy that there isn't much structure behind it. Kaizen fits perfectly with the principles of Concept Management.[16]

Just-in-Time (JIT)

Just-in-time (JIT) is the American name given to the Toyota production system that focused on continuous improvement and waste elimination through the use of Kanbans (inventory tracking devices). It has been successfully copied in many plants throughout the U.S. The benefit of the JIT process is reduced-cost production; the limitation of the system is restricted product flexibility. The JIT process has been adapted throughout industry, retailing, wholesale distribution, and, as we saw in Chapter 1 in the case of Doyle Wilson, construction.

Toyota, however, has eliminated the restrictions of traditional JIT as used in the U.S. Toyota is able to produce any vehicle in any sequence, totally customized within 4 hours. And the flexibility does not come at the expense of efficiency; Toyota is still able to totally turn its inventory once every 3 hours, which over two shifts is 1000 turns per year.

JIT efficiencies are worthy of study and careful understanding. This is an invaluable tool for change efficiency improvements.[17]

Process Mapping (Flow Charting)

Process mapping (often referred to as *flow charting*) is a tool that can be applied in many different ways. For example, if we want to map the accounts receivable process, we would tape all the documents, notes, papers, and procedures on a big wall, showing flow lines between all processes and identifying all lines of communication. Then an analysis process would be made of each of the steps to see if they were value-adding or a waste, making that step now able to be improved for efficiency. Specific benefits of process mapping include

- Identify opportunities
- Identify the scope of the change project
- Analyze the current process
- Define objective outcomes
- Identify root causes of problems
- Test and refine measurable results

Process Reengineering would take a different approach. It would recommend looking only at the end and redesigning the process with the end in mind. We would pretend that no process existed, and we would rethink the objective of the process.

Breakthrough Thinking would take an even broader approach, asking the question "What is the purpose of the accounts receivable system?" Through the purpose hierarchy, we may decide that there is a better approach than the accounts receivable process to achieve our objective. Breakthrough Thinking should be an essential part of any process mapping project.

Quality Circles

The concept of quality circles (QC) was a first attempt at copying the ideas of the Japanese. QC was installed to implement teaming focused on quality improvement. These teaming circles have taken on a broader approach through TQM, where they are focused on change management and innovation in all areas, including quality. The basic concept still exists, but it has now incorporated the broader approach required by a change-management system.[18]

Statistical Process Control (SPC)

Statistical process control (SPC) is an analysis tool that focuses on process control. It uses statistical methods of sampling to measure and monitor performance results in an attempt to reduce and improve the quality process at the same time. SPC catches potential quality problems before they occur. It is a valuable tool that complements the performance of the TQM process.[19]

> *The world is moving so fast these days that the man who says it can't be done is generally interrupted by someone doing it.*
>
> **Harry Emerson Fosdick (1878-1968)**

Quality Gurus

Since Japan has developed such a strong competitive stance over the last decade based on quality, a collection of quality experts have taken the stage, each with their own program designed to assure quality success.

Deming became popular for his statistical process control programs in Japan. The Deming Award in Japan was named after him. Juran has ten steps to quality improvement, from which he developed over 70 years of quality improvement effort.[20]

Crosby has designed a program focused on quality that is less precisely defined but is nevertheless more understandable than other programs.

Each of these quality experts has programs that can be incorporated into a TQM implementation and can be focused on quality improvements. Many of the steps of their programs, such as setting goals, training, and organizing, are also a key element of a TQM process.[21]

Manufacturing Resources Planning (MRP II)

Manufacturing resources planning (MRP II) is the integration of the production, logistics, financial, marketing, customers, vendors, and engineering functions into one integrated database of information. This allows accessibility to engineering drawings, customer sales, etc., from anywhere. The integration of all these information systems offers the user an excellent tool for TQM analysis and development.[22]

Enterprise Resources Planning (ERP)

Enterprise resources planning (ERP) is an outgrowth and expansion of MRP II and activity-based costing. ERP uses MRP II's integrated database to evaluate the resource efficiency of the entire organization. It incorporates *finite capacity scheduling* to create a real-time enhancement to the scheduling process. Once again, this is an excellent tool for the evaluation and analysis of the process and resource efficiencies of an organization through the TQM process.[23]

Time-to-Market and Cycle Time Management

Time-to-market (TTM) and cycle time management focus on cutting the lead time for new product development. We want to get new ideas out and available to the customers quicker than the competitors. They focus on issues such as

- Reduced wasted (non-value-added) resources
- Reduced rushes into and within the process
- Increased process effectiveness, efficiency, and adaptability
- Reduced Crosby's "re" world — rework, retype, rethink, etc.
- Changes in what and how we do things

With TTM tools, we can use TQM to become more responsive to those changes that are the most critical to customer satisfaction.[24]

Supply Chain Management

Supply chain management is a recent buzzword that focuses on the ability to manage the entire network, from vendors through the logistics process (shipping and warehousing) through the factory and out the door to the customer. The focus is to achieve cycle-time reductions and increase quality. This is again an excellent tool to use in conjunction with TQM to focus on supplier to customer efficiency improvements.[25]

Endnotes

1. Kobu, Bulent and Frank Greenwood, "Continuous Improvement in a Competitive Global Economy," *Production and Inventory Management*, Fourth Quarter, 1991, pp. 58-63.
2. There are many good sources for a discussion of Total Quality Management (TQM) and the other techniques discussed in this chapter. The primary sources for the discussion of these techniques in the book are: Plenert, Gerhard, *World Class Manager*, Prima Publishing, Rocklin, CA, 1995; Hoffherr, Glen D., John W. Moran, and Gerald Nadler, *Breakthrough Thinking in Total Quality Management*, PTR Prentice Hall, Englewood Cliffs, NJ, 1994; Omachonu, Vincent K. and Joel E. Ross, *Principles of Total Quality*, St. Lucie Press, Deray Beach, FL, 1994; Hammer, Michael and James Champy, *Reengineering the Corporation*, Harper Press, New York, 1993; Ross, Joel E., *Total Quality Management* (Second Edition), St. Lucie Press, Delray Beach, FL, 1995; Mahoney, Francis X. and Carl G. Thor, *The TQM Trilogy— Using ISO 9000, The Deming Prize, and the Baldrige Award to*

Establish a System for Total Quality Management, AMACOM, New York, 1994; Bhote, Keki R., *World Class Quality*, AMACOM, New York, 1991.
3. Many of the ideas in this section are taken from the book by Plenert, *World Class Manager*, where the ideas are elaborated on in more detail.
4. Bemkowski, Karen, "The Quality Glossary," *Quality Progress*, February 1992, pp. 19–29.
5. See the references to teaming in the endnotes for Chapter 3.
6. Review the discussion of TQM in Plenert's Book *World Class Manager* to get dozens of references.
7. The following publications and additional information are available from AT&T's Customer Information Center, Order Entry Department, P.O. Box 19901, Indianapolis, IN 46219; 1-800-432-6600. AT&T Bell Laboratories, *AT&T's Total Quality Approach*, Publication Center of AT&T Bell Laboratories, 1992; AT&T Bell Laboratories, *AT&T Process Quality Management & Improvement Guidelines*, Publication Center of AT&T Bell Laboratories, 1989.
8. Plenert, Gerhard and Shozo Hibino, "The T-Model: A Systematic Model for Change," *National Productivity Review*, Vol. 13, No. 4, Autumn 1994, pp. 543–549.
9. Plenert, Gerhard, "Process Re-Engineering: The Latest Fad Toward Failure," *APICS - The Performance Advantage*, June 1994, pp. 22–24.
10. This quote, the reengineering principles, and the Three-R's come from Hammer, M. and J. Champy, *Reengineering the Corporation*, Harper Business, New York, 1993.
11. A more detailed comparison of TQM and Process Reengineering can be found in Plenert's book *World Class Manager*.
12. Hammer, M. and Champy, "The Promise of Reengineering," *Fortune*, May 3, 1993, pp. 94-97; Hammer, M., "Reengineering Work: Don't Automate, Obliterate," *Harvard Business Review*, July-August 1990, pp. 104–112; Jason, R., "How Reengineering Transforms Organizations to Satisfy Customers," *National Productivity Review*, Winter 1992, pp. 45–53; Marks, Peter, *Process Reengineeering and the New Manufacturing Enterprise Wheel: 15 Processes for Competitive Advantage*, CASA/SME Technical Forum, Society of Manufacturing Engineers (SME), Dearborn, MI, 1994; Harbour, Jerry L., *The Process Reengineering Workbook: Practical Steps to Working Faster and Smarter Through Process Improvement*, Quality Resources, White Plains, New York, 1994; Cypress, Harold L., "Re-engineering," *OR/MS Today*, February 1994, pp. 18–29; Ravikumar, Ravi, "Business Process Reengineering — Making the Transition," *APICS 37th Annual International Conference Proceedings*, APICS, October 1994, pp. 17–21; Miller, George, "Reengineering: 40 U$seful Hints," *APICS 37th Annual International Conference Proceedings*, APICS, October 1994, pp. 22–26; Melnyk, Steven A. and William R. Wassweiler, "Business Process Reengineering: Understanding the Process, Responding to the Right Needs," *APICS 37th Annual International Conference Proceedings*, APICS, October 1994, pp. 115–120; Boyer, John E., "Reengineering Office Processes," *APICS 37th Annual*

International Conference Proceedings, APICS, October 1994, pp. 522–526; Stevens, Mark, "Reengineering the Manufacturing Company: 'New Fad or For Real'" *"APICS 37th Annual International Conference Proceedings*, APICS, October 1994, pp. 527–530.
13. Some excellent articles that focus on the QFD procedure are: Stocker, Gregg D., "Quality Functional Deployment: Listening to the Voice of the Customer," *APICS 34th International Conference Proceedings*, APICS, October 1991, pp. 258-262; Henrickson, Dave, "Product Design as a Team Sport," *Target*, Spring 1990, pp. 4-12; In the Wallace book there are two sections worth looking at: Wallace, Thomas F., *World Class Manufacturing*, Oliver Wright Publications, Essex Junction, VT, 1994; I-3 "Linking Customer to Strategies via Quality Functional Deployment (QFD)" by Thomas F. Wallace; II-6 Quality Functional Deployment: Breakthrough Tool for Product Development" by William Barnard.
14. The Wallace book also has an article by Robert L. Jones and Joseph R. Tunner titled "ISO 9000: The International Standard for Quality." ISO information can be obtained from any of the quality and productivity organizations. The ISO organization, International Organization for Standardization, is located in Geneva, Switzerland.
15. The Wallace book listed earlier has an article titled "The Malcolm Baldrige National Quality Award Program" by Stephen George. Malcolm Baldrige application information can be obtained from: Malcolm Baldrige National Quality Award, National Institute of Standards and Technology, Gaithersburg, MD 20899; (301) 975-2036. Information about the Deming Prize criteria and the Baldrige Award criteria can be obtained from the book *Breakthrough Thinking in Total Quality Management* listed earlier in the Endnotes of Chapter 1. The Shingo Prize for Excellence in Manufacturing criteria can be obtained from College of Business, Utah State University, Logan, UT 84322-3521; (800) 472-9965; http://www.usu.edu/~shingo
16. Because all the references to kaizen are short and philosophical, it becomes difficult to recommend additional reading. Perhaps the book *Principles of Total Quality* by Omachonu and Ross mentioned earlier will be helpful.
17. The Japanese production systems are discussed in numerous books. A thorough discussion of the JIT process can be found in the book Plenert, Gerhard, *International Management and Production: Survival Techniques for Corporate America*, Tab Professional and Reference Books, Blue Ridge Sumit, PA, 1990. Numerous books are available that detail the Toyota production version of JIT: Shingo, Shigeo, *Study of the Toyota Production System from the Industrial Engineering Viewpoint*, Japanese Management Association, Tokyo, 1981. Shingo has worked with Toyota and has an insider's viewpoint. Also, Wantuck, Kenneth A., *Just in Time for America*, The Forum, Ltd., Milwaukee, WI 1989.
18. Quality circles are discussed in the Omachonu and Ross book listed earlier.
19. SPC is discussed in detail in Bhote, Keki R., *World Class Quality*, American Management Association, New York, 1991; Johnson, Perry L. , Fred K. Miller, Jon C. Plew, and Marcia A. Sikora, *Easy as SPC: A Programmed-*

Instruction Workbook for Statistical Process Control, Perry Johnson, Inc., Southfield, MI, 1986.
20. Juran's ten steps to quality improvement are found on page 11 of the book *Breakthrough Thinking in Total Quality Management* listed earlier in the Endnotes of Chapter 1.
21. An excellent comparison of the Deming, Juran, and Crosby focus points can be found in Mahoney, Francis X. and Carl G. Thor, *The TQM Trilogy: Using ISO 9000, the Deming Prize, and the Baldrige Award to Establish a System for Total Quality Management*, American Management Association, New York, 1994, pp. 133–137.
22. MRP II is discussed in detail in the book *The Plant Operations Handbook*, listed earlier in the Endnotes of Chapter 3.
23. The best source of information on ERP is through the recent publications of the American Production and Inventory Control Society (APICS). You can get a catalog by phoning (800) 444-2742.
24. Chapter 7 of the book *World Class Manager* focuses on TTM and has a long list of additional references.
25. The American Production and Inventory Control Society (APICS) listed earlier, as well as the Council of Logistics Management (CLM) in Oarbrook, IL, (708) 574-0985, both have excellent publications on supply chain management efficiencies.

CONCEPT MANAGEMENT

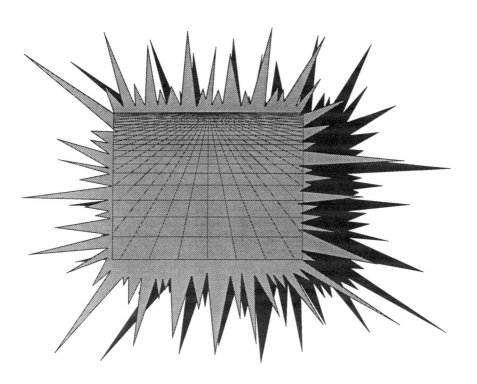

Chapter 5

Concept Management Philosophy

Consider the diagram. This is a connect-the-dots diagram that I want you to do: Start with any dot and connect all the dots with one continuous line to see a picture.

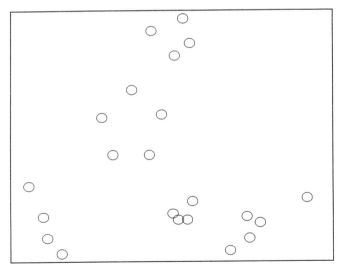

Did you get a good image? Do you know what the picture should look like? If not, why not? Running an organization is often like connecting a collection of loose dots together. We're not sure what the final result should look like. We have no projection of what the future has in store for us. We simply connect dots because we were told to, and the more dots we connect the better managers we are. But what if we were given a goal; what if we were told that the dots should be connected to form a sailing ship? Would that change how we connected the dots? Of course it would! Go back, knowing that the dots form a sailboat, and attempt again to connect the dots. Do you see the sailboat?

What if you had not just the dots and the information that the dots form a sailboat, but you were also given a roadmap of how to connect the dots? This time attempt to connect the dots in the next diagram and see what you come out with.

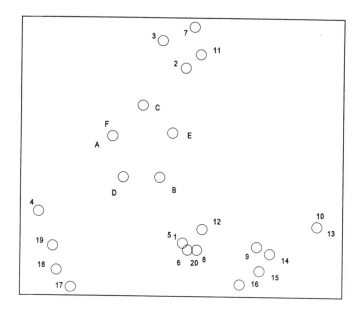

This time the dots should have come out looking very much like a sailing vessel. It should look like the drawing to the right.

Concept Management is identifying your goals by defining your future (the knowledge that we want a sailing ship), and then by developing a system of guiding steps, a roadmap (the sequentially numbered dots) to help us achieve our objective.

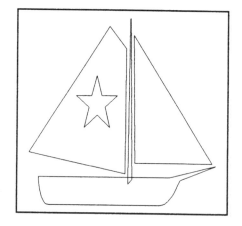

Concept Management Philosophy

*That which is beautiful is not always good;
but that which is good is always beautiful.*

***Sliders*, The Professor**

Everyone is familiar with the Walkman radio by Sony Corporation. People all over the world enjoy listening to music as they exercise, work, or play. The Walkman was invented in 1979 by Mr. Masaru Ibuka, one of the founders of Sony Corporation. One day, after his retirement, he was walking around his factory, and he encountered a developmental team that was working on the development of ultra-micro recording technology. As they worked, they were enjoying background music. Mr. Ibuka had the inspiration that this technology should be integrated with the micro headphone technology that was being worked on in an entirely separate department. He received an enormous amount of resistance to the idea; opponents claimed that no one would want a tape player that could not record. However, Mr. Ibuka persisted, and together with Mr. Morita developed the nonrecording tape-player with headphones that we today know as the "Walkman."[1] There was no market for a walking music player; however, through the creative breakthrough process, a new product was developed that created not only a market, but an entire culture. The breakthrough process was used, the change function was initiated, and the result was a newer, higher plain of success.

Concept Management works in a series of stages. These stages are all part of the whole, which is Concept Management. The stages are as follows:

1. Concept Creation — The development and creation of new ideas through the use of Breakthrough Thinking's innovative methods of creativity.
2. Concept Focus — The development of a target, which includes keeping your organization focused on core values and a core competency. Then, utilizing the creativity generated by Concept Creation, a set of targets are established using World Class Management, and a road map is developed helping us to achieve the targets.
3. Concept Engineering — This is the engineering of the ideas, converting the fuzzy concepts into usable, consumer-oriented ideas. TQM through the use of a focused, chartered team and through a managed systematic problem-solving process helps us to manage the concept from idea to product.

4. Concept In — This is the process of creating a market for the new concept, similar to the Sony Walkman experience. We transform the concept into a product, service, or system, using World Class Management techniques. We may utilize Breakthrough Thinking to help us develop a meaningful and effective market strategy.
5. Concept Management — Both the management of the new concepts as well as a change in the management approach (management style) is affected by the new concept. Concept Management is the integration of the first four stages of the Concept Management process (Creation, Focus, Engineering, and In).

Let's go on to further understand what Concept Management is all about.

Concept

In discussing the stages of Concept Management we first need to understand two terms: Concept and Management. "Concept," according to Webster's Third International Dictionary, is defined as "... a general or abstract idea, ... a generic mental image. It is a general mental image abstracted from precepts."[2] There are two parts in this definition. The first is "a mental image" and the other is "precepts" or viewpoints. From this we develop our definition of "Concept" as being "the mental image based on an integrated viewpoint." From the perspective of Breakthrough Thinking this becomes

1. An integrated viewpoint — purposes, values, measures
2. A mental image — ideal target image which projects a solution

Therefore, the new concept starts with a redefinition of purposes, values, and measures to create new images, which in turn project redefined purposes, values, and measures. A structural (graphical) representation of "Concept" can be seen in the next diagram.

In this diagram, we start with a redefined purpose.[3] This redefined purpose presents us with an ideal goal that we will attempt to shoot for. We take this redefined goal and revalidate it against our core values. Then, with this redefined purpose (goal) we can redefine our measures and objectives through World Class Management techniques and develop a strategy for successful implementation.

Concept Management Philosophy

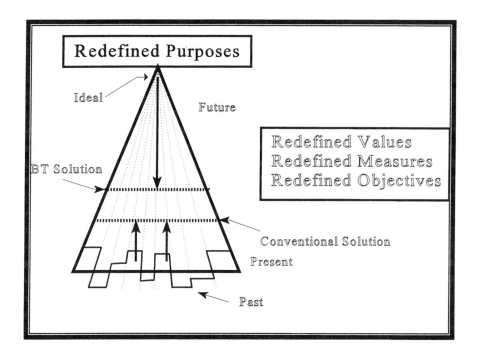

From the earlier diagram we see that conventional (root cause) analysis would focus on the past and the present to move forward toward the future. Some systems are still in the past, some are in the present, and some are in the future. However, they are all at different stages and they are not necessarily integrated together with a focus on the ideal. Utilizing the analytical approach would move us forward toward the future slowly, one small step at a time. This forward-moving solution moves you slowly, at a snail's pace, toward the ideal target. Concept Management starts with the ideal solution as developed through purpose expansion, and moves us backward toward reality, always searching for the first, most idealistic solution that is implementable. The BT solution requires creativity and a leap of faith, but it jumps us ahead of the busywork of putting out fires that were built based on present or past techniques, and this movement occurs one small step at a time.

It is critical to note the gap between the conventional solution and the BT solution. This gap defines competitiveness. It defines the difference between where the competition will find itself, and where you will find yourself. Concept Management will help you leap-frog into the future, significantly ahead of your competition.

Management

The second part of the term "Concept Management" that requires further analysis is the word "Management." The use of the term follows the definition offered by *World Class Manager* which stresses the following characteristics required of a World Class Manager:[4]

Sunrise Manager — Having a long-term orientation; looking for a "better way"

Theory-Z Manager — Having employees involved with and guiding the business process through participative and empowered team efforts

Change Manager — Guiding a dynamic, evolving business organism that capitalizes on change opportunities

Leader — Being a character-building, motivational example

We need World Class Leaders to effectively manage a Concept Management culture and organization.

Concept Management

Understanding the terms "Concept" and "Manager," we can now visualize what is expected of a Concept Manager (CM). In the next diagram we see a graphical representation of how Concept Management is the integration of the three key subsystems, Breakthrough Thinking, World Class Management, and Total Quality Management. The purpose is defined by Breakthrough Thinking (BT). This is then converted into definable and measurable goals through the use of World Class Management (WCM). World Class Management offers tools and ideas. One of these tools is Total Quality Management (TQM), which establishes a structural, consistent approach for implementing change. TQM offers the control dimension of CM. In the diagram we see a focus and movement toward the purpose through the use of WCM tools, ideas, values, and measures, and all this utilizes TQM as the planning, control, and feedback mechanism.

Concept Management principles were utilized in the development of 3M's Post-it Notes, where a simple in-house idea was turned into a major corporate business sector. It was used by Federal Express, where the vision of one individual created a revolution in the shipping industry. It was used at Toshiba, which has established a Department of Concept Engineering within its Research and Development organization that innovated the Dyna-Book

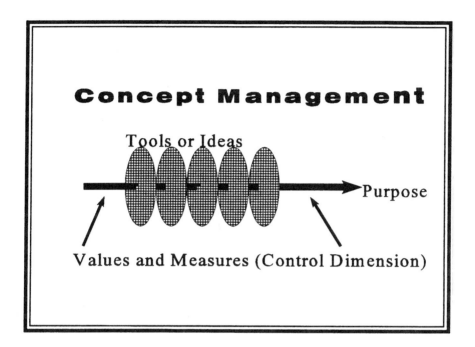

Word Processing system. It was utilized by Fuji Camera when it created the disposable camera, creating an entirely new industry sector. In each of these examples, we can see creativity leading to a purpose, which gives focus and a mission. The creativity is then engineered, marketed, and managed. In the diagram at the top of the next page we see Concept Management as it would apply to organization-wide innovation.

Concept Management principles, when applied to specific problem-solving situations, are diagramed at the bottom of the next page.

These diagrams give us a larger vision of how Concept Management principles are integrated. Next we will discuss a badly needed shift in the way we think. Then we will focus on a discussion of each of the Concept Management principles.

A Paradigm Shift in Thinking

Concept Management builds on a new paradigm, a paradigm shift, away from conventional "push" style thinking, to a new, more creative "pull" style. Traditional, conventional-style thinking focuses on "root cause analysis." For example, Ford Motor Company has developed an elaborate

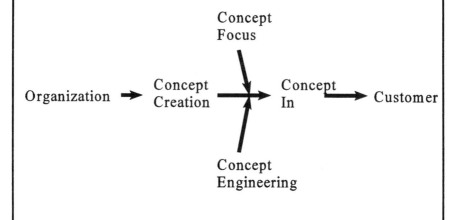

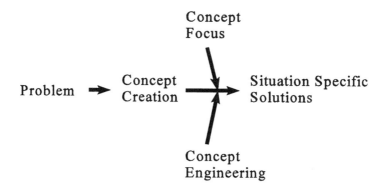

Concept Management Philosophy   **89**

system that focuses on isolating problems and identifying their root causes. Recently, TRIZ has been adapted from Russia, and is another, similar form of root cause analysis. Traditional Total Quality Control systems have focused on data collection with tools like Statistical Process Control in an attempt to identify root causes.

In "push" style thinking, we assume that a study of the past will identify the problems of the future. We "push" conventional, traditional thinking into the future. We project very little change, assuming that the past is a fair representation of what is to come. For example, in our business schools we utilize "case studies" to study how to do business correctly. Someone once said that "as long as American business schools are studying how the Japanese did business 10, 15, or even 20 years ago, the Japanese will have no difficulty in defeating them in the future." Case studies incorrectly assume that the future of Japanese business will simply be a projection of how business was done in the past. Another example of push thinking is our use of forecasting and financial analysis techniques. Conventional forecasting says that we take a history of past performance and project similar, consistent behavior out into the future; the future is the same as the past. In financial analysis, we generate financial analysis numbers 20, 30, and even more years out into the future, assuming that the past is an adequate predictor of future profitability, sales, interest rates, government regulations, and even customer interests. **The basic fallacy of PUSH thinking is that it assumes that the past is an accurate representation of the future; it assumes that *CHANGE WILL NOT OCCUR !***

In the following diagram we see how in the past we followed a certain path, assuming that the path is continuous and will lead us directly into the future. However, reality indicates change. Had they been ready for change, National Cash Register Corporation (NCR) would not have been so stubborn about shifting from mechanical to electronic cash registers, insisting that the consumers want trustworthy mechanical devices. From this mistake, International Business Machines (IBM) was born and NCR no longer exists. Similarly, if IBM had been ready for change, IBM would not have been so blind about the advent of the small computer, resulting in their decline and the impressive growth of Intel and Microsoft. Now IBM is on the decline. If American automotive manufacturers would not have insisted that they were smarter than the consumer and that the American consumer did not want small cars, the Japanese automakers would not now control such a large share of the American automotive market. And the stories go on and on.

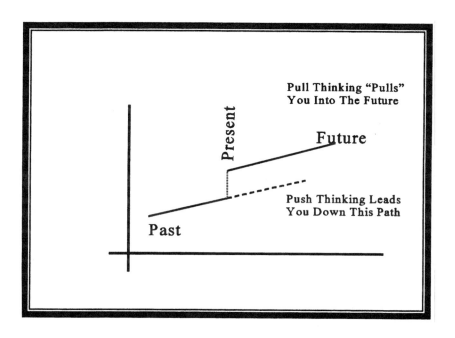

Reconsidering the diagram, we see how conventional "push" thinking would tell us that the past can be pushed into our understanding of the future. However, the only thing sure in life is change! Pull thinking looks at the future, plans to achieve the future, and then builds our corporate strategy around this future. We, through our goals and strategy, will then pull our organizations into this future vision of ourselves.

Understanding the difference between push and pull thinking, we can see how the management of a company is offered several strategies:

Product-out — We produce a product and push it out on the marketplace. Traditional "make-to-stock" manufacturing.

Market-In — The market pulls our products from us through the production process. Just-in-time (JIT) production planning as utilized by Toyota.

Seeds-Out — Research and Development (R&D) defines a new product and pushes it out to the consumer.

Concept-In — Research and Development is not an entity in and of itself; rather, it is part of an integrated team of cross-functional people including marketeers, engineers, production people, customers, vendors, etc. This integrated team focuses on people (all stakeholders in the company and its products), purposes (from the Breakthrough Thinking hierarchy), and values (from World Class Management's core principles). Concept-In is not the marketing of an engineered product, it is the integration of an entire selection of needs that often creates new products and then goes forward to market those products.

In the age of mass production, the product-out strategy was extremely successful. For example, Henry Ford had a better idea with his assembly line, which utilized Concept Creation. His products were "product-out" pushed into the marketplace. The cost benefits of this strategy made the process successful. However, we have progressed to a "market-in" era where consumers are used to having options and enjoy being highly selective. Customers want customized products and markets. They need to be able to drive what is being produced. That is why the Toyota JIT system has been so successful. It is a pull, market-driven system.

Traditional R&D techniques assume that we should establish research think-tanks that have the freedom to create. That type of R&D program is a "seeds-out" push strategy that develops a lot of technology, but very little of it actually generates consumer products. The United States spends a lot of R&D money through its research laboratories. A much better strategy is the "concept-in" strategy where the creative effort is teamed with cross-functional areas such as marketing, thereby maintaining a focus in the creative process.

Concept Creation

Concept Creation focuses on the development and creation of new ideas through the use of Breakthrough Thinking's innovative methods of creativity. In earlier chapters we learned how Breakthrough Thinking breaks away from traditional analytical (root cause) problem analysis. Rather, Breakthrough Thinking focuses on purpose expansion. The methodology of utilizing purpose expansion for problem solving is not a process that occurs only at one particular stage of the management process. Rather, it occurs at numerous stages of the process. Concept Creation occurs at the early stages, for example, when the corporate vision and mission are being developed. Concept Creation again occurs when we are focusing on strategy development. And later, as teams have been organized which focus on specific charters, Concept Creation will be utilized to identify opportunities for improvement, replacement, or even elimination.

Concept Creation takes many forms. For example,

Customer Concept Creation — Concept creation utilizing customer ideas

Time-To-Market Concept Creation — Concept Creation focusing on time-to-market performance

Regardless of the specific focus, concepts are created through the purpose expansion process; then they are focused, engineered, marketed, and managed.

Concept Focus

Concept Focus is the development of focus. What should the focus of our effort be directed toward? How do we define and identify that focus? How do we measure and motivate our employees toward a common, organization-wide focus?

Concept Focus is the development of a target, which includes keeping your organization focused on core values and a core competency through the use of World Class Management planning processes (see Chapter 3 and the book *World Class Manager*). Utilizing the creativity generated by Concept Creation, a set of targets are established, and a road map is developed helping us to achieve the targets. Concept Focus is the guiding light that keeps all the team planning processes focused on the same unified target.

Concept Focus starts by establishing a cultural shift within our organization much the way Ricardo Semler did at Semco (see Chapter 1). All members of the organization need to see themselves as an integral part of the decision-making process. They need to feel the freedom to make and change organizational decisions. They need to feel that their opinions have value and that management cares about them personally. They need to see managers as leaders and facilitators that help them accomplish their goals and implement their ideas, rather than judge them. They need to move out of the mode of feeling like an organizational tool, and move into a mode of feeling that they are the organization.

Concept Focus is the first basic stage of change within the organization. This process requires an organization-wide cultural shift toward change. It is possibly the most difficult stage, since it requires evidence. Employees won't believe the culture of the organization has changed just because you told them so. Management's actions and attitudes are the evidence necessary to demonstrate that this change is actually happening. Don't be surprised if this process takes a year or more.

Concept Engineering

It is impressive how, a few years ago, small and relatively unknown organizations such as Microsoft and Apple, with their weak and unimpressive Research and Development organizations, were each able to outdo IBM with its four big Research and Development institutes. A small organization with properly created and properly focused concepts has no difficulty outmaneuvering the giants who have become wrapped up in bureaucracy. Engineering without focus is a waste. We repeatedly see small companies start out with creativity, then become bogged down with their superiority, loosing sight of the vision and getting wrapped up in the bureaucracy, and then come crashing down. We saw it with NCR Corporation, we see it with Bell Labs and IBM, and we will see it in the future with Microsoft and Intel.

All good ideas need development. As with the Sony Walkman, Masaru Ibuka came up with the idea (Concept Creation) which fit within the focus of the company (small electronics products — Concept Focus). Next, the new concept required development into a consumer product. This is the engineering of the ideas, converting the fuzzy concepts into usable, consumer-oriented ideas. This is done by taking the idea and turning it over to a Concept Engineering team using TQM teaming principles. The team must be chartered (focus), motivated (gainsharing), and given the authority to bring the concept into reality (empowered). TQM through the use of focused, chartered teams and through a managed SPS process helps us to manage the concept from idea to product.

Concept In

Will customers come after a better mouse trap if they don't know that it exists? Marketing is the realization and creation of needs in customers. Concept In is the taking of the engineered product, identification of the market, and development of a "pull" marketing strategy based on the new product (refer to the discussion earlier in this chapter under "a paradigm shift in thinking"). The product we transformed from concept into a product, service or system, using World Class Management techniques is now applied using Breakthrough Thinking tools to develop a marketing strategy. It is wise to retain team consistency through this process. For example, the team that did the engineering should also be the team that does the Concept In process. In TQM, we learned that the teams should be cross-functional, and therefore, an engineering team should have contained marketing people, and a marketing team should contain engineers. Therefore, the same cross-functional team that develops the concept into a product should also be responsible for developing customer excitement for the new product.

An example of the Concept Management process can be seen with Microsoft Corporation. Microsoft created a new software concept and proposed it to the world. They wanted to develop user-friendly computers so that customers would feel comfortable having computers in their homes. They developed the Windows user interface. The concept creation was customer (user) friendly computers. The Concept Focus was an expanded customer market. The Concept Engineering was the development of the Windows product; and the Concept In was the marketing of the new Windows product. Windows required a new type of customer. The bit-picking computer techies were happy in their distinguished and isolated world of DOS. They didn't have a need for a higher-level user interface such as Windows. Microsoft had to introduce and teach the nontechie user about this new interface. It had to create a market for its Windows product. Today, Windows has become the standard of the personal computer industry, used by techies and nontechies alike.

Concept Management Philosophy   **95**

Concept Management

Here is a problem:

> Concept Creation — Create a symmetrical diagram that takes ten points (or coins) and arranges them so that five straight lines can cross all ten points with each line crossing four points.
> Concept Focus — Solve another frustrating and irritating problem.
> Concept Engineering — Develop the symmetrical diagram that fits the conceptual requirements. Hint: you've already encountered this diagram earlier in this book.
> Concept In — Identify a customer market for crazy and frustrating problems. Maybe you can get someone else to solve the problem for you.
> Concept Management — Go to Appendix 5A at the end of this chapter and look up the answer.

Concept Management is both the management of the new concepts as well as a change in the management approach (management style) toward change management, team-focused empowerment, and World Class Management techniques. Concept Management is the integration of all the stages of the Concept Management process.

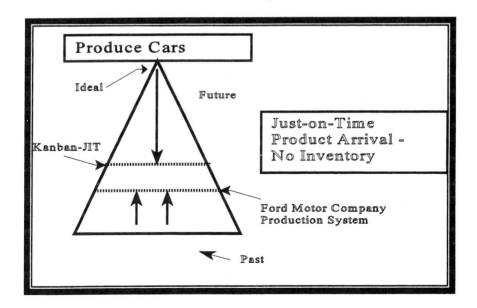

Sakichi Toyoda, CEO of the Toyota Weaving Company, had a son, Kiichiro Toyoda, who wanted to venture out and create his own mark, much like Ricardo Semler of Semco. Kiichiro wanted to start a car production facility. He separated himself from the Toyota Weaving Company and founded Toyota Motors. He hired Taiichi Ohmo from the Nagoya Institute of Technology and stressed that in post-World War II Japan, financial and resource availability pressures required a focus on inventory minimization. Automobiles should only be produced to satisfy a customer demand, never for inventory. And the components for the production process should arrive just on time for the needed process to occur. In other words, there should be no finished goods inventory (only build to customer orders), there should be no work-in-process inventory (materials arrive just on time for the process and are immediately utilized), and there should be no raw materials inventory (just on time arrivals).

A trip had been scheduled to visit Ford Motor Company in the United States in order to learn how to produce automobiles. Ohmo insisted that he wanted to go along. He was fascinated by the concept of supermarkets. He had read in the newspaper that American supermarkets were able to minimize inventory levels and have high inventory turns. He was fascinated by the idea that such a large business such as a supermarket could have such low inventory levels. Supermarkets did not exist in Japan, and Ohmo wanted to go on the Ford plant visit so that he could see what made the supermarket so successful.

When the Toyota Motor company delegation arrived in Michigan, Taiichi Ohmo disappeared. The remainder of the Toyota representatives visited and studied the Ford assembly line process. However, Ohmo was nowhere to be found. At the end of the visit, Ohmo reappeared and was questioned about his disappearance. He explained to the Toyota directors that he already knew how Ford worked and that their inventory levels were too high. He wanted to see how the supermarket had lower inventories. He claimed that he had learned the secret. The secret was money. Money was the source of the inventory exchange transaction. No money, no exchange.

Taiichi Ohmo convinced Kiichiro Toyoda that a production process should require a device similar to money for the exchange of inventory and that this would minimize inventory levels. From this concept came the birth of what we today know as the "Kanban" inventory control devices. Kanbans (cards) are used like money to get materials. Kanbans drive purchases and production; no Kanban, no materials. Only if you have a Kanban are you given materials. Kanban has become a control device that works very similar to money in a supermarket. Out of this simple beginning came the production management philosophy known today as Just-in-Time (JIT) production management.

Concept Management Philosophy 97

Looking at the JIT development process through the eyes of Concept Management, we see

> Concept Creation — Produce cars without inventory; a search for innovative techniques.
> Concept Focus — Car Production
> Concept Engineering — Ford Motor Company processes could not be copied. A new process had to be engineered: Kanban-based JIT.
> Concept In — Sell the idea to Kiichiro Toyoda
> Concept Management — The development of a new concept for automobile production that fit within the restricted materials resources and financial needs of post-World War II Japan. Rather than imitation being allowed to limit the creative process (conventional thinking), Breakthrough Thinking was utilized to create a concept and a focus that went far beyond an analysis of present or past processes.

The concept management process did not end with the development of JIT and Kanbans. Rather, this was just the beginning. The change culture had been initiated and was flowing freely. For example, the development of work design techniques (waste elimination), or the Andon System (system of indicator lights) were further concept creation ideas that maintained the concept focus perspective.

Summary

Concept Management is an organization-wide culture change refocusing the organization on *change* and on the integration of a series of processes:

- Concept Creation
- Concept Focus
- Concept Engineering
- Concept In

Concept Management utilizes a series of tools to accomplish the integration. The tools it uses are

- Breakthrough Thinking — The creativity driver.
- World Class Management — The organizational and integration driver.
- Total Quality Management — The implementation process.

In the next chapter we will proceed to reorganize ourselves from top to bottom toward a Concept Managed competitive organization. It will teach us how to Concept Manage our organization.

Appendix 5A

The answer is easy, once you understand the symmetry. Here's the diagram you should come up with (remember the sail on the sailboat — that was your hint):

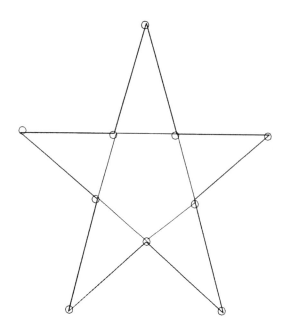

ENDNOTES:

1. Takeuchi, G. and S. Nojima, *Breakthroughs*, Diamond Publishing, pp. 50–51.
2. Gove, Philip B., Ed., *Webster's Third International Dictionary*, Merriam-Webster, 1970, p. 469.
3. "Breakthrough Thinking," *Participant Handbook,* The Center for Breakthrough Thinking, 1986, p. 41.
4. Plenert, Gerhard, *World Class Manager*, Prima Publishing, Rocklin, CA, 1995, pp. 13–14.

Chapter 6

Concept Management — Integration and Organization: "How CM Works"

Wisdom is realizing that you never know everything there is to know about anything. Never assume you know what you're doing! Challenge everything!

Gerhard Plenert

Three men rent a hotel room for the night. They pay $30, each man paying $10. A little later the hotel manager realizes that he overcharged for the room, that the price of the room should have been $25. He sends the bellboy to the room with $5 to repay the visitors. The bellboy pays each man $1 and keeps the remaining $2 for a tip. Now, in the end, each of the men paid $9 for the room, making a total of $27. If we add the amount that we paid the bellboy ($2) we get a total of $29. But initially we had paid $30. What happened to the other $1?

This question is a classic example of misdirected attention. We are working on the wrong problem. The distraction is keeping us from seeing the real problem (check Appendix 6A for an explanation of this problem). That is how organizational change is often approached. We attack the symptoms, and not the problem. In our development of Concept Management in this chapter, we will be establishing a change management system that is clearly focused and properly structured.

> A World Class Person is not the things we do; the things we do are the tools to our becoming World Class.

When we look for improvements/changes/continuous improvement, we are often picking the low-hanging fruit near the bottom of the tree. What we need to reach for is the fruit at the top of the tree, which is riper and sweeter. It will take additional imagination (Breakthrough Thinking) and often it isn't easy to reach. You'll need a little imagination, but the results are so much better.

KAO of Japan utilized Concept Creation (Breakthrough Thinking) techniques to develop a principle that they refer to as the *regularity concept*. In this concept they look at the six central standard deviations of the normal curve and evaluate performance in this region. Then they develop systems that focus on

1. The most important factors or items
2. The six sigma region of performance

They ignore the outliers in that they refuse to build systems that are oriented solely on solving outlier situations or problems. For example, KAO applied this concept to its accounting functions. One of the processes in this area was the reconciliation of travel expenses. KAO felt that the majority of its employees were honest, and that it was generating a tremendous amount of waste by having all its employees file detailed travel expense statements. This required the collection of receipts, the documentation of trips, and the filling out of extensive forms. KAO felt that, since most employees were honest, why should all these forms be necessary? It decided to implement a system wherein only individuals who had a history of transgression were subject to the reconciliation process. After 1 year, KAO found that even these audits were unnecessary, since most of the transgressions were misunderstandings. Therefore, it eliminated the reconciliation process entirely. When employees return from a trip, they input their expenses directly into a computer, and within minutes the expenses for their trip are reimbursed directly into their bank account.

Concept Management — Integration and Organization: "How CM Works" 101

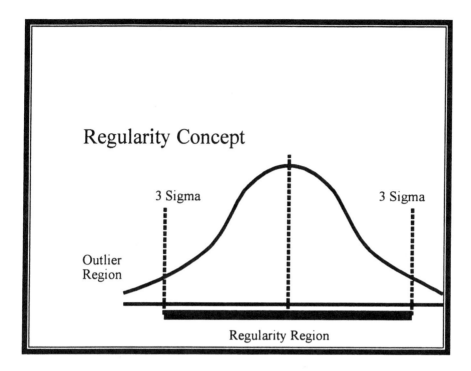

KAO took the *regularity concept* and implemented it throughout its accounting department. In the end, a department that originally contained a staff of 150 employees was reduced to 7 employees.

What makes the KAO reimbursement program so interesting is that the U.S. Defense Department of the Navy spent $2 billion during 1995 on travel expenses, and $2.2 billion on reconciling travel expenses. We, as a nation, would have been better off if we would have allowed everyone in the Navy to take their spouses with then on their trips.

At the beginning of Part II of this book we have a diagram. In the center of the diagram we see a vision that reaches out into the future. Concept Management reaches out toward future competitiveness. However, at the same time, Concept Management is an explosion out of the box searching for a future of more creative, competitive change.

This chapter will utilize the elements of Concept Management discussed in the last chapter:

- Concept Creation
- Concept Focus
- Concept Engineering
- Concept In
- Concept Management

We will then proceed to define some of the steps an organization should go through to fully implement the organization-wide cultural change necessary to make Concept Management an organization-wide success.

Making Innovation Happen: Concept Management Through Integration

In the amusement centers throughout the world they have a game where you take a hammer and take out your frustrations on groundhogs. The groundhogs pop up through a series of holes. As they appear through the hole, the player attempts to hit them with the hammer as quickly as possible, the quicker the hit, the more points are earned. However, as you hit the one groundhog, another appears, and the skillful player is the one who hits the most groundhogs as quickly as possible in rapid succession.

Traditional Descartes approaches to problem solving are similar to hammering groundhogs. We are constantly fighting to put out fires. A good manager is one who can put out the most fires the quickest. No one ever steps back and asks; "Why are we hitting the groundhogs

Concept Management — Integration and Organization: "How CM Works"

(fighting the fires)?" Concept Management stresses the importance of asking questions about which game we are playing. For example, root cause analysis techniques focus us on groundhogs. We identify the root cause of one problem and attack that root cause, not realizing that as soon as we hammer that problem, another groundhog quickly rears its head.

Process Reengineering, like so many of today's management tools, focuses on squeezing the employees. We treat our employees as towels that need squeezing. Once upon a time, these towels were very wet. For example, when the largest value-added component of product cost was labor, our towel was very wet. There was a lot of squeezing that could be done. However, today, in nearly all types of manufacturing, the value-added component of labor is less than 10%, and often less than 5%. Materials, in most cases, are more than 50%, and if we add to materials cost the amount of carrying cost of the materials (normally hidden in burden or overhead), then the materials value-added costs often exceed 70%. In most production processes, the burden cost is a larger percentage of the value-added cost than is the labor cost. However, we continue squeezing the labor cost towel and ignore all the other wet towels that are lying around, such as inventory cost, facilities cost, burden cost, maintenance cost, etc. And, as a by-product, by continually squeezing the labor towel through programs such as downsizing, we are generating employee dissatisfaction and distrust, the exact opposite of what is needed in a World Class setting.

We've discussed it in pieces, and now it is time to integrate Concept Management into one integrated process. Returning to our philosophy of how change is implemented in Chapter 1, we define target objectives (ideal solutions) and draw a road map:

- Concept Creation utilizing Breakthrough Thinking, and through Purpose Expansion, helps develop creative, innovative goals (target solutions). Breakthrough Thinking and World Class Management visioning offers a new vision of the road map's purpose, and offers a methodology that will make the trip shorter, with fewer bumps.
- Concept Focus through the use of World Class Management helps formalize those goals into a structure that outlines the focus and format of the target. World Class Management offers the structure to design the road map. It locates where we are now and helps us build a map toward our target solution.
- Concept Engineering through Total Quality Management focuses on the employees through the teaming process. It standardizes how teams should be organized and chartered. It formalizes the change implementation process.
- Concept In utilizes World Class Management and Total Quality Management to generate marketing processes.

So what does all this mean? How do you actually implement Concept Management within your organization? There is nothing the authors hate more than reading a book filled with a lot of good-sounding but very fuzzy ideas. We don't want to leave you fuzzy about Concept Management, even though there is a very strong philosophical element to this concept, as you discovered in the last chapter (Chapter 5). Concept Management works! And thus we will now discuss the formalization of this process through a series of steps:

1. Top Management Commitment — Or a least a commitment at the highest level possible. However, there are some elements, such as a change in the measurement/motivation system that cannot be changed without this top-most level commitment.

2. Training — Thinking paradigm tools such as Breakthrough Thinking, TQM, and World Class personal development cannot be achieved through osmosis. They must be taught. They are not difficult to teach, but the training is critical; without it failure is guaranteed.
3. Goal expansion through Breakthrough Thinking — Use Breakthrough Thinking to identify the Values and Core Competencies as defined by World Class Management. Identify a focus purpose and from this focus purpose create a vision, mission, and strategy for your organization. This strategy is then the key in identifying areas for TQM teams.
4. Quality Council — A top-level team needs to be established that defines charters (empowerment, gainsharing, measurement/motivation) and identifies the purposes of the teams that are to be organized.
5. Corporate-Wide Purpose Expansion — Build a purpose expansion chart wherein every employee at every function lists all his or her tasks, identifies the purpose of the task, identifies the purpose of the purpose, and then classifies the function as important (✔), unimportant (✘), or undecided (?). Proceed with the immediate elimination of all unimportant tasks.
6. Team Establishment and Chartering — The toughest part of this project is to be believable to the employees. You need to demonstrate your trustworthiness. Don't be surprised if this step takes over 1 year. However, the objective in this step is to establish empowered, gainsharing, self-motivated, self-directed, goal-focused teams that will address a charter and attack this charter with viciousness and drive.
7. Utilize Systematic Problem Solving, as established by the Quality Council, to standardize the organization's change process.
8. Establish a philosophy of continual change and continual training, including the reading and team discussion of books such as *The Goal*, *Maverick*, *Small is Beautiful*, *Breakthrough Thinking*, and *World Class Management*.[1]

Now, we will work through this process in detail.

Top Management Commitment

Top management commitment must follow the World Class Management principles of leadership. Top management must be trained in Breakthrough

Thinking, World Class Management, and Total Quality Management techniques. Top management must be committed to the creative process, to empowerment and gainsharing, and to teaming. Top management must recognize the line between measurement and motivation and must be willing to adapt the measurement process in the same way that all processes need to be changed. This is critical in motivating the proper responses and employee performances.

Top management commitment is more than just hanging signs on the wall. It requires management to consciously attempt to affect the culture of the organization. A cultural shift is required of all its members. Some of the key areas of this cultural shift include:

1. Establishing common goals that everyone shares in, understands, and has an influence on and a responsibility in achieving. The organization needs a set of beliefs, behaviors, assumptions, and a vision that everyone shares in and believes in. Goals require ownership, not just verbalization. The goals should be focused and measured through a control and feedback mechanism.
2. The organization needs a change orientation. Status quo or tradition can never be an adequate excuse for doing anything. The organization's culture should realize that success requires competitiveness, and competitiveness requires change.
3. A systems perspective is needed where we always focus on the ideal. Root cause analysis techniques focus backward. Creativity techniques focus forward.
4. A participative, team-oriented, Theory-Z, World Class management style is required where employees are empowered to make changes, chartered so that they know what to change, and receive benefits for the changes made (gainsharing).
5. A redefinition of the corporate structure is needed. A power shift downward toward the team is critical. Management's role should be that of facilitator, not judge and jury. Management is responsible for making things happen that were decided on by the teams. It is not management's role to pass judgment on team decisions. This requires an enormous amount of trust. Managers need to trust their employees, and employees need to trust that management won't criticize them or fire them for their ideas.
6. As Ricardo Semler states, the best policy is no policy. We need to avoid the over-bureaucratization of our organizations. Too many rules, policies, and procedures stifle creativity.
7. Organize around results, not around fires. Help the organization to think idealistically and long-term. Avoid getting wrapped up in the day-to-day activities.

Our recommendation is that top management, from the Board of Directors through the Vice President level, should all receive the recommended training. As part of this training, they need to be involved in visioning and organizational focus development. They should use Concept Creation to define Core Competencies and Core Values for the organization. Then they should utilize Concept Focusing to develop their perspective on the Vision Statement and the Mission Statement for their organization.

Training

Concept Management is a collection of tools and thinking paradigms. Not all the tools are appropriate for all situations. Some tools may never be used. But you need to know about the tools in order to effectively and intelligently make decisions about their appropriateness and usefulness. Concept training is necessary at all levels of the organization for the basic tools of Concept Management: Breakthrough Thinking, World Class Management, and Total Quality Management.

Our recommendation is to initiate an overview, half-day program that would introduce Concept Management to everyone in the organization. This should be followed by a question/answer session that allows everyone to get involved in the process and to express their opinions. We want all employees to identify with the process and to buy into it (have ownership in it). After the overview, an initial test program should be started, selecting the organization that is the most enthusiastic about the process. These employees will become the salesmen for the rest of the organization. This initial test program should focus on a detailed understanding of the basic tools of Concept Management. This test process should generate some interesting success stories that can then be shared throughout the organization. Following the test process, similar training should occur throughout the organizations so that all employees can become involved in the process.

Goal Expansion

Concept Creation utilizes Breakthrough Thinking as its driving engine. Through the purpose expansion process (BT) and creative visioning (WCM), the ideal essence and purpose of the organization can be explored. Then, utilizing World Class Management, these goals can be given structure and focus. The sequence of steps that would be followed are something like the following:

1. Top management and the ownership of the organization would utilize the BT paradigm, BT techniques, and WCM visioning to define the core values and the core competencies of the organization.
2. Groups of employees at all levels of the organization are also asked to generate their own versions of the core values and core competencies.
3. Consensus cores statements are agreed upon.
4. All levels of the organization are asked to work on a vision and mission statement for the organization.
5. Consensus vision and mission statements are developed.
6. From the mission statement, specific corporate goals and milestones are established; for example, a certain percentage of market share, or a specific level of quality. These milestones are then analyzed to identify the critically contributing resources (which resources have the greatest effect on the success of this goal). This will be the root of the measurement and motivation systems that will become the charters of the TQM teams.
7. A corporate strategy is developed sufficient to cover the requirements of the mission statement (about 10 years).
8. A business unit strategy is developed that expands on the corporate strategy but at the business unit level.
9. Corporate and business unit operating plans are developed.

An excellent book that lists 301 corporate mission statements from America's top companies is *The Mission Statement Book*.[2] These mission statement suggestions should by no means be considered to be the best, or even good, but you can get some very useful examples and ideas from this book.

Our recommendation is that the outlined procedure be followed so that a structured set of goals and guidelines can be established. This will give the organization the required guidance and direction.

Quality Council

A top-level cross-functional team, usually at the CEO and VP level, needs to be established. This team becomes the organizing spearhead for the entire organization and takes on a lot of the CEO functions. However, the perspective of the council is different from the traditional CEO role. The council sees the organization as a system, made up of a series of subsystems. Each subsystem has resources (employees, materials, finances, etc.) and produces some objective output. Systems can overlap and share

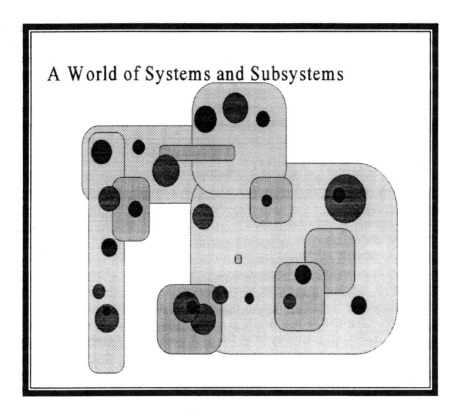

resources and responsibilities. However, each subsystem is identifiable, unique, and changeable.

It is the role of the council to identify systems, to assign cross-functional teams of employees to work on those systems, to charter the systems, and to measure and motivate the systems. For example, the types of systems analyzed can include

    Process systems (process teams)
        Production process
        Accounts receivable process
        Shipping process
        Engineering process
        Etc.
    Product systems (product teams)
        A specific model of a car
        Customer service
        Customer maintenance
        Etc.

Project systems (project teams)
A new building construction
A new production line installation
A new computer system installation
Etc.

The Quality Council defines charters (empowerment, gainsharing, measurement/motivation) that identify the purposes of the teams that are to be organized. It sets up an appropriate measurement system that will motivate the team. It monitors the performance of the teams by establishing a Systematic Problem Solving procedure that standardizes the change process for each of the teams, thus allowing the council to easily check the performance of any of the teams.

So what does the CEO do? The CEO, like all other managers, is now a facilitator. The CEO interfaces between the quality council and the Board of Directors or other external stakeholders. And the CEO communicates the concerns, desires, and wishes of the board back to the council. An authoritarian CEO does not fit the World Class model.

Our recommendation is to establish a Quality Council as soon as possible in the process. Council members need to become comfortable with their new roles as interfaces and facilitators. For example, the financial VP would now be on the Quality Council on which he represents the views of the finance department, and would also sit on the financial department management team, acting as the facilitator of the team's needs. The financial manager may also sit on other teams throughout the organization as cross-functional needs require representation. With the new Quality Council in place, the power reins need to slowly be turned over to them.

Corporate-Wide Purpose Expansion

Mitsubishi Electric (Panasonic), an organization with 60,000 employees, started a Breakthrough Thinking training program in 1991. By 1993 they had accomplished a major restructuring of the corporation. One of the programs they found to be the most helpful was the use of a corporate-wide purpose expansion program. They found the following:

Necessary	52.77%
Unknown	28.93%
Unnecessary	18.30%

Slightly more than 50% of the tasks that they were engaged in were actually needed for the success of the organization. The other half were either not needed or their value was unknown. The organization was able to engage in an enormous work reduction and job elimination program, which focused them on value-added tasks. But that did not mean layoffs, or else the employees would not have cooperated to begin with. It meant that new value-added tasks and functions could be created for employees as "waste" tasks were eliminated. Mitsubishi Electric found this exercise interesting because the employees were so eager to eliminate the "waste" tasks. Employees felt an ownership in the change process, because they were the ones who identified what tasks should be eliminated. This adaptation of the "people design concept" energized a creativity that was unexpected by Mitsubishi's top management.

Corporate-wide purpose expansion is utilized to flush out the garbage. A lot of what goes on in an organization is busywork and does not contribute value added to the organization. Busywork is a waste. Corporate-wide purpose expansion reviews the tasks of each employee and attempts to eliminate the waste tasks.

To do corporate-wide purpose expansion we must first have all employees trained in Concept Creation (Breakthrough Thinking) principles. Then we proceed by having all employees, from the CEO down to the shop floor sweeper, develop a purpose analysis chart. The chart has every employee, at every function, list all their tasks, then identify the purpose of the task, and then identify the purpose of the purpose, as shown on the chart. Next the employee classifies each purpose as important (✔), unimportant (✘), or undecided (?).

Task	Purpose	Purpose of Purpose	Rank
Filing	Data collection	History	✘
Cut parts	Add value to parts	Prepare product for customer sale	✔

Once the purpose expansion has been completed for each of the tasks, the unimportant tasks should immediately be eliminated.

As the outside world is always changing, people have to change correspondingly. People have to change their tasks, often in a dramatic and drastic way. They have to create new tasks and eliminate the old tasks, which creates changes to all tasks. Our function should never be to keep the present task, but to creatively look for ways to destroy it.

Our recommendation is to do the corporate-wide purpose expansion as soon as possible after Concept Creation training. However, the employees must have a sense of the new corporate culture or they will not tend to identify tasks as unnecessary. The reasoning is that if an employee identifies much of their work as unnecessary, they fear that they will be eliminated. They need to understand that they will not be eliminated for performance improvements. That is not World Class! World Class stresses that all employees will always remain employable, either within the organization or outside of it (the organization will assist the employees in retraining and finding new jobs). Therefore, our recommendation is to proceed with the corporate-wide purpose expansion, making sure that the cultural shift for the organization is in place.

Teams

The Quality Council develops teams, charters the teams (focus and direction), empowers them, and applies gainsharing (team members receive benefit for their performance). As already discussed under the corporate-wide purpose expansion program, the toughest part of this project is to be believable to the employees. The teaming process should follow the TQM methodology discussed in Chapter 4. The synergy of team creativity has been well established and is part of a World Class working environment.

Our recommendation is that the entire organization should think in terms of team performance and team results. Individual contributions should be recognized within the team structure, but corporate recognition should occur only at the team level, thereby encouraging team synergy. For example, an award can be given to the team that has demonstrated the best performance. However, the distribution and use of the reward should be left up to the team. Perhaps they want to throw a big party for everyone, or perhaps they want to give the entire reward to one person. This element of the reward process should be a team decision. However, the corporate perspective should be focused on team contributions in order to encourage synergy as much as possible.

Systematic Problem Solving (SPS)

Systematic Problem Solving is a basic element of an effective TQM system. It is established by the Quality Council and standardizes the organization's change process. All change is implemented through a series of steps. By identifying which step a project is on, and by reviewing the feedback mechanisms of that step, it is easy to tell the status of any project within an organization. The book *World Class Manager* discusses three SPS processes that have been very effective: the one used by Florida Power and Light, the one used by AT&T, and a new technique called the T-Model. None of these techniques is best; however, they all should be considered in the development of your own customized SPS procedure.

Our recommendation is that you establish the SPS process immediately and make it part of the training that occurs early in the program. SPS, like everything else, should be considered to be dynamic and should be receptive to change.

Philosophy of Continual Change

All things are open to change, including the newly established corporate philosophy on change. Establish a philosophy of continual change and continual training, including the reading and team discussion of books such as *The Goal, Maverick, Small is Beautiful, Breakthrough Thinking, The Seven Habits of Highly Effective People*, and *World Class Management*.[3] Individuals should be encouraged to be as creative and imaginative as possible. Never should any idea be rejected, no matter how inappropriate you feel it is. Discouraging bad ideas will often discourage future good ideas. All ideas should be open for consideration; never should any idea be ridiculed.

> *The mode of thinking in organizations — not only about TQM, but in all problem-solving situations — needs to change.*
>
> **Hoffherr, Moran, & Nadler,
> Breakthrough Thinking in Total
> Quality Management, page 45.**

The Flow of it All

It is helpful to have a chart listing all the steps discussed in this chapter. You should take this list, modify it to fit your needs, and hang it on the wall. Then work through the process. Recognize that this is not a step-by-step process. Some of these steps need to be accomplished simultaneously. However, this list of steps is a starting point for future planning.

1. Top Management Commitment
    A. Common goals
    B Change orientation
    C. Systems perspective
    D. World Class management style
    E. Redefinition of the corporate structure
    F. The best policy is no policy
    G. Organize around results
2. Training
3. Goal Expansion
    a. Define the core values and the core competencies
    b. Employee versions
    c. Consensus cores statements
    d. Multiple vision and mission statements
    e. Consensus vision and mission statements
    f. Goals and milestones are established
    g. Corporate strategy
    h. Business unit strategy
    i. Corporate and business unit operating plans
4. Quality Council
    Process Systems
    Product Systems
    Project Systems
5. Corporate-Wide Purpose Expansion
6. Teams
7. Systematic Problem Solving (SPS)
8. Philosophy of Continual Change

Increased competition, especially from foreign sources, has made the hotel market extremely competitive in Japan. Many hotels, and even entire chains, have closed their doors because of bankruptcy. This competitive threat inspired the Hotel Nagoya Castle chain to focus on concept management techniques and to move away from traditional management approaches. They utilized TQM and World Class Management teaming and Breakthrough Thinking training to develop new organizational concepts. One of their initial investigations into the operation of the hotel had them do a purpose expansion of the hotel. It went something like this:

What is the purpose of a hotel?
   To provide accommodation for one or more nights
   To provide accommodation for relaxation
   To change your environment
   To change your mood
How do you get people to come back to the same hotel and still change the environment and mood?
   Make each visit feel like a new, exciting experience
   Make each floor of the hotel a different environment

The purpose expansion process is repeated and continued on similar to the above. Finally, Nagoya decided that it would establish each floor of the hotel to represent a different mood and environment. There is a red floor, a blue floor, etc. (the red floor gets the highest level of demand and is always sold out). The furniture, fixtures, wall hangings, etc., all reflect the mood of the floor. The floors each have a special feeling; quietness, excitement, etc. Additionally, the hotel made a special attempt to respect the privacy of the visitors. As a result of this strategy, they were able to increase their repeat customer base so much so that they did not find themselves in the same position that the other local five star hotels found themselves in — bankruptcy.

The Hotel Nagoya Castle, the headquarters hotel, was a traditional five-star hotel, but competition with hotels like the Hilton had put them on the verge of bankruptcy. They installed a major management change incorporating the newly learned management techniques. New programs were installed to increase competitiveness. For example, in an attempt to generate increased interest in the hotel, they tried to involve the community in their 40-year anniversary celebration. One way of encouraging this involvement was to have an art exhibition. Utilizing conventional thinking, they decided to look at a similar exhibition that was arranged by the Tokyo Hotel. This approach had been to invite local artists to display their art in the hotel and then to welcome visitors to view the art pieces. They sold tickets to recover the costs. The ticket sales were subcontracted out to ticket sellers. However, after considering the Tokyo Hotel approach, someone suggested a purpose expansion be performed. The purpose expansion went something like this:

What is the purpose of an art exhibit?
   To show gratitude to the community
   To enjoy art
   To invite people to share in the 40-year anniversary celebration
What is the purpose of inviting people to share in the 40-year anniversary?
   To celebrate the 40-year anniversary
   To get continuing community support for the hotel
   To have a future

The purpose expansion continued on like this in the hope that all the alternative purposes would be discovered. Then they looked for a focus purpose, which is the highest-level purpose that is achievable. The focus purpose was:

To give thanks to the community
To gain community support for the hotel
To have a future

They came to the realization that an art exhibition had too much cost associated with it (advertisement and facilities arrangements) and did not attract enough attention in the community. They decided that an art festival would be much more effective. In an art festival, the local art museums would be invited to come and display their art pieces in the hotel. The art museums were interested in selling the pieces and would be allowed to advertise. The art museums also handled all the advertisement, most of which came free through the local news media in the newspapers and television. Since the art festival was free, a much greater number of people came to the festival than would have come to a fee-charged exhibit (this follows the principle of Concept In — pulling sales). There was no ticket handling or selling — pushing sales. The festival was an overwhelming success because it was unique and free. About 35,000 people attended. Sales were great for the art museums, and for the hotel's restaurants and other facilities. Community exposure occurred, and the art dealers, the hotel, and the community (because of the free tickets) all came away with a very positive feeling.

What's Next?

Over the next few chapters we will break the Concept Management process down a little further. In the next chapter we will focus on the role of measurement and motivation. This relationship becomes very important in Concept Management, and it is often a difficult one for us to get a handle on. Then we will discuss people, both inside and outside of the organization, and how the synergy of teaming works. As a wrap-up, we have a Concept Management test to see if your thinking approaches have become World Class, or if you still have a long way to go.

Summary

This chapter has given you a series of steps for the implementation of Concept Management. These steps are not "hard and fast" procedures for implementation. The key to successful implementation is flexibility and a willingness to change, including the change of how Concept Management is utilized.

A man was working on his car. He was lying under the car when his wife happened to walk by and he called out to her, "Could you hand me a wrench?"

His wife said "sure" and proceeded to hand him a screwdriver.

The man, irritated by the mistake, said "How can someone so beautiful be so dumb?"

His wife responded, "God made me beautiful so that you would marry me, and He made me dumb so that I would marry you."

> *The guidance we need ... cannot be found in science or technology, the value of which utterly depends on the ends they serve; but it can still be found in the traditional wisdom of mankind.*
>
> **E. F. Schumacher, *Small is Beautiful***

Appendix 6A

Here is the way we should have viewed this problem. The room cost $25 Adding to that the $2 paid to the bellboy gives us $27, which is equal to 3 times the $9 that each man paid. If we want to get back to the $30, we need to add the $1 that each man paid and later received back from the bellboy ($3 total) to the $27.

Endnotes

1. Plenert, Gerhard, *World Class Manager*, Prima Publishing, Rocklin, CA, 1995; Nadler, Gerald and Shozo Hibino, *Breakthrough Thinking*, Prima Publishing, Rocklin, CA, 1990; Nadler, Gerald and Shozo Hibino with John Farrell, *Creative Solution Finding*, Prima Publishing, Rocklin, CA, 1994; Hoffherr, Glen D., John W. Moran, and Gerald Nadler, *Breakthrough Thinking in Total Quality Management*, PTR Prentice Hall, Englewood Cliffs, NJ, 1994; Semler, Ricardo, *Maverick*, Warner Books, New York, 1993; Goldratt, Eliyahu M. and Jeff Cox, *The Goal*, North River Press, Croton-on-Hudson, New York, 1986. Schuhmacher, E. F., *Small is Beautiful*, Perennial Library, New York, 1973.
2. Abrahams, Jeffrey, *The Mission Statement Book: 301 Corporate Statements from America's Top Companies,* Ten Speed Press, Berkeley, California, 1996.
3. Plenert, Gerhard, *World Class Manager*, Prima Publishing, Rocklin, CA, 1995; Covey, Stephen R., *The Seven Habits of Highly Effective People*, Simon & Schuster, New York, 1989; Nadler, Gerald and Shozo Hibino, *Breakthrough Thinking*, Prima Publishing, Rocklin, CA, 1990; Nadler, Gerald and Shozo Hibino with John Farrell, *Creative Solution Finding*, Prima Publishing, Rocklin, CA, 1994; Hoffherr, Glen D., John W. Moran, and Gerald Nadler, *Breakthrough Thinking in Total Quality Management*, PTR Prentice Hall, Englewood Cliffs, NJ, 1994; Semler, Ricardo, *Maverick*, Warner Books, New York, 1993; Goldratt, Eliyahu M. and Jeff Cox, *The Goal*, North River Press, Croton-on-Hudson, New York, 1986; Schuhmacher, E. F., *Small is Beautiful*, Perennial Library, New York, 1973.

Chapter 7

Concept Management: Measurement and Motivation

> *No matter how much legislation the American Civil Liberties Union (ACLU) promotes through our legal systems and our courts, there will never be an end to school prayer as long as there are finals.*

A young boy was told how important he was to the family and how the family was counting on him. He was told that he made a valuable contribution in the family and that without him the family couldn't be whole. They especially counted on him to help and take care of his younger brother by not fighting with him and trying to make him happy.

The young boy thought it over for a minute and then he said, "Can you count on someone else?"[1]

Do we look for someone else to count on when we get tough challenges? Maybe the young boy wasn't motivated sufficiently to perform better in this area. You may wonder why we've decided to dedicate an entire chapter to measurement and motivation. The reason is simple: it's the area in which the most mistakes are made. We could tell you the story of the plant manager who attempted to follow "lean manufacturing" principles (run the plant with no wasted resources) only to nearly get fired because inventory reductions had destroyed current ratios. The following month he purchased all the inventory back. Or we could tell you the story of the plant that had one of the best quality systems available: they had statistical process control (SPC), total quality control (TQC), TQM, banners, quality control (QC) circles, and more. However, the plant was being closed because of poor quality. We were invited to this plant to discuss why quality was so poor in spite of the fact that the installed systems were so elaborate and leading-edge. The answer was simple: they were measuring performance and paying bonuses based on throughput through the work center. The evaluation systems stressed throughput, and no matter how much hype was given to quality, throughput made the difference between getting fired and getting more pay.

If you want quality, you have to measure quality. If you want operating cost reduction, that's what you need to measure. Don't pretend that measuring performance in one area will increase performance in another. Reflecting back on what we have discussed in pieces in several of the past chapters, we need to follow these steps:

1. Identify the goals.
2. Identify the critical resource — the resource that has the most significant effect on achieving the goal. For example, in automotive manufacturing, if the goal is profitability, the value-added contribution numbers come out approximately as follows: labor — less than 10%; materials — 50–60%; machinery — less than 10%; and overhead — 20–30%. In this example, the critical resource, materials, has the most significant contribution to profitability and should be the resource around which we plan efficiencies and improvements (review Chapter 3 for more details).
3. Develop a measurement and motivation system around the critical resource. Remember that measurement systems exist for purposes of motivation, not for data collection.

As we stressed earlier, improving the efficiency of the wrong resource can actually hurt profitability. Therefore, it is important to realize that *increasing quality does not ensure increased competitiveness, it may decrease it! Increasing productivity does not guarantee increased profits, it may actually decrease profitability!*

Intuitively, this does not seem correct, because our entire training has told us that increasing productivity and quality is a good thing! However, increasing productivity and quality in the wrong areas reduces the performance efficiency in other areas. Increasing labor efficiency in the automotive example above will reduce inventory efficiency, resulting in lost profitability.

Concept Management teaches a few principles about the measurement–motivation relationship that we need to discuss in more detail. We will now review a few of these concepts and then we will integrate them into the overall Concept Management framework.

Customer Quality

In Chapter 3 we learned about customer satisfaction under the World Class model. To reiterate, *a satisfied customer is one who is so excited about your product or service that they wouldn't think about purchasing it from anyone but you.*

Although this definition is good, it is not the best. The synergy generated by Concept Management tells us that this view of the satisfied customer is still based on conventional thinking. The first problem with the definition is that it only attempts to "satisfy" the customer. This means that the customer "thinks about," likes, and purchases our product because they were happy with it. However, what we really want to do is to "thrill" our customer to the point where they are emotionally attached to it. We don't want them to do any more "thinking about" which product to purchase, we want them emotionally excited about and committed to our product. The ultimate purpose of our product should be that everyone who encounters it is delighted. Therefore, the Concept Management objective is not customer satisfaction, but *customer delight. Delighted customers* are so excited by and attached to our product that they are automatically and unconsciously attracted to it. *Excited customers* are at the point where they will select our product based on emotion, and not simply based on logic.

The ultimate purpose of developing and selling a product is to create customer delight, not just customer satisfaction. If customers are delighted, they will be emotionally driven to change to our product, and to remain loyal and committed to it after the change. This is the Concept Management level to which we want to bring all our customers.

We want to measure and motivate customer delight. To do this we need to interact with the customers and find out what delights them. The product team, responsible for a specific product, needs to find out what it is that generates this delight. Satisfaction is not sufficient. And the team also needs to realize that the target "delight" changes over time. It can't be discovered once and then be expected to remain the same. It is a constantly changing target.

An excellent book on quality is *In Search of Quality,* which lists opinions on what quality should be of 43 industrial leaders, including the

CEOs of world class companies and quality leaders such as Deming, Juran, and Crosby.[2] These statements should not be considered the last word on quality. Rather, they should be considered a benchmark we should all be ready to beat.

Productivity Thinking and Value-Added Thinking

One important new operational measure of Concept Management is "thinking productivity." Concept Management focuses on the ability to *do the right things, before you do things right.* Another way to express this is by stressing *think smarter, not harder.*[3]

Thinking productivity challenges the way we do things. It questions the purpose of all activities. For example, the purpose of data collection is not for cost accounting, but rather for motivation. If data collection is not generating the appropriate response from the factory and its resources, then it is a waste.

Thinking productivity looks closely at the improvements generated by change. Is change a regular part of your organization? Are the changes generating positive responses from your employees, from your customers, on your operational measures of performance, etc.?

Thinking productivity also ties in closely with the concept of value-adding. Thinking (creativity) that is not goal-directed and purpose driven is wasted thinking. Even within the thinking process, we should identify a purpose. What is the purpose of thinking about the problem? What is the purpose of solving the problem? Then, by evaluating the purpose hierarchy we may quickly realize that the analytical thinking process we were about to engage in is a waste (non-value-adding).

Absolute Benchmarking

Benchmarking takes on a new twist in Concept Management. It becomes more than just a performance comparison tool. It becomes a purpose-driven, goal-focused opportunity to become the best. An excellent way to diagram absolute benchmarking is with a triangle, as shown in the next diagram.

Conventional benchmarking would drive us toward the comparative benchmark goal. We want to get better than before, or we want to get better than our competition. This type of thinking focuses on present and past status. However, *absolute benchmarking* would drive us toward a higher, future, ultimate objective. Perhaps this is best illustrated by a story.

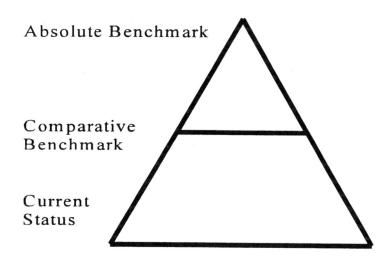

Toyota focuses on inventory efficiency. It sees its absolute benchmark as the elimination of all inventory. It is not satisfied with the comparative benchmark of being more inventory efficient than before, or of being more inventory efficient than its competition. Toyota sees its goal as the absolute elimination of all inventory, and has effectively achieved it with having only 3 hours of inventory in the plant at any one time. When we realize that it takes 4 hours to produce a car from start to finish (which is also quite impressive), then we realize that they don't even have enough inventory to produce one car in the factory at any one time.

Another twist to the Toyota story involves the inventory control information system. Conventional benchmarking would suggest that we look at our inventory control system and try to make it better, or to evaluate the competitor's inventory control system and to make ours better than theirs. However, absolute benchmarking suggests that we need no inventory control system at all if we have no inventory. And that is the approach that Toyota has taken — they have no inventory control system.

How Should We Measure?

The best way to discuss what should be measured is to reflect back on the discussion in Chapter 5. We need to define, redefine, and re-redefine our measurement system as the value system changes. We need to focus the measurement system on the redefined organizational purpose. Putting this concept in World Class Management terminology, we would suggest the following:

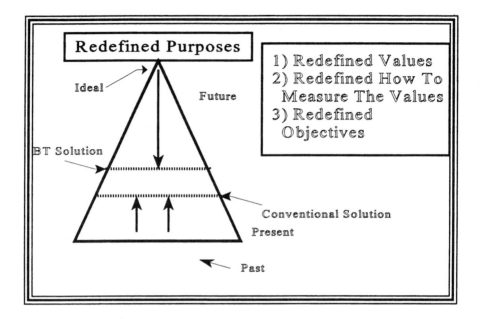

Your ideal purpose is your vision statement.
Your focus purpose is your mission statement.
Your values and objectives are your organizational and business unit strategy.

The benefit that concept creation (Breakthrough Thinking) brings to concept focus (World Class Management) is that it projects us to a higher set of values and objectives, ones that will have a more global (ideal) perspective.

Comparing the traditional measurement methodologies to Concept Management measurement can be seen in an example. If we are to measure our performance in writing a letter we could measure it under traditional methodologies that would focus on the number of errors as follows:

Purpose — write letter
Value — correctly written letter
Measure — percentage of errors
Objective — less than 5% of errors

A quality letter would be one that has less than 5% errors. The 5% mark would be the measure of success.

Recognizing the uniqueness principle of Chapter 2, Concept Management would measure the quality of the letter in more subjective terms. It would suggest a five-point evaluation system where

A = excellent
B = good
C = average
D = below average
E = poor

Using this rating system, each letter would be subjectively evaluated as follows:

Purpose — write letter
Value — beautifully (excitingly) written letter
Measure — five-point evaluation system
Objective — A or B rating

What Should We Measure?

The objective of Concept Management measurement systems can be summarized by the statement, *focus on the purpose, not on the numbers!*

Conventional methods focus on measuring the "speed of accomplishment." Concept Management focuses on the "effectiveness of the accomplishment" or on the "long-term value of the accomplishment."

> *Resources flow toward what is measured.*
>
> **Tom Tuttle**

If we are measuring short-term improvements, our resources, including the human resource and its efforts, will focus on improvements in short term. Therefore, if we want long-term improvements, customer quality, etc., then we need long-term measures. The U.S. automotive industry has still not converted its plants to JIT (just-in-time) production methods, even though experiments with plants such as Saturn have been overwhelming operational success stories. This change hasn't occurred because everyone from the CEO down is measured on short-term performance measures, and a conversion to JIT is an extremely long-term project. With such performance measurement, who cares if it is beneficial to the consumer (higher levels of quality and lower costs of operation in JIT), to the long-term stability and even existence of the company (competition), etc. What's most important is that everyone, from the CEO on down, satisfy the needs of the measurement system so that they can "keep their job and get a raise."

Trenton Forging Company of Trenton, Missouri was struggling with the problem of identifying methods of increasing grinding productivity in their forging area. They had been working on this problem for quite some time when, as part of their ongoing training, they received Breakthrough Thinking training. After working through a purpose hierarchy for the grinding problem they soon realized that not only should they not be working on increasing grinding productivity, they shouldn't be doing grinding in the forging area at all, a realization that hadn't even occurred to them earlier. They were so focused on the measurement system, increasing productivity, that they lost sight of the big picture, improved organizational performance and profitability. The Breakthrough Thinking process was so effective for them that they utilized it to become the first forging company in the world to be QS-9000 certified. During this certification process, a period of about 1 year, as a side benefit of being purpose-focused, they increased quality, increased sales by 25%, and increased their technical sophistication to make them more competitive.[4]

Summary

In the Gobi desert there is a regiment of troops whose function it is to keep the tracks of a railroad clear of sand. The purpose of the railroad is to deliver food and supplies to the troops. Does this strike you as strange? Or are you comfortable doing things just because they've always been done a certain way. Like soldiers in the Gobi desert? Traditional measurement systems are not only ineffective, they are usually a waste, and they are often destructive.

IBM's new president has recognized the relationship between measurement and motivation. As part of the corporate restructuring, he has established new measurement standards that promote customer responsiveness. IBM now realizes that customer relationships build profits. It also realizes that products need to be designed to satisfy customer needs, rather than pushing engineering-created products on the customers. Similarly, UPS has changed the measurement system for its delivery employees. Rather then measuring how many packages are delivered per hour, it now measures repeat customer performance. UPS has come to the realization that you get repeat customers if the delivery employees become friends of the customers, not if they just rush in and out of the customer's site in a hurry.

This has been a short chapter, but it is a critical chapter. The organizational culture needs to change, and a key element in the successful implementation of this change is to develop proper measurement systems. Employees won't believe that a culture change has occurred if "things remain the same" in the area of measurement. We need to break free from traditional hard number methodologies, and focus on the uniqueness principle. We need to utilize Concept Management measurement approaches in order to become a World Class organization.

Endnotes

1. Adapted from a talk by J. Christensen, October 5, 1996.
2. Shelton, Ken, *In Search of Quality; 4 Unique Perspectives, 43 Different Voices*, Executive Excellence Publishing, Provo, UT, 1995.
3. "Think Smarter, Not Harder" and many other concepts, catch phrases, and slogans in this book, such as the "Paradigm Shift in Thinking," are registered trademarks of The Center for Breakthrough Thinking, Inc., University of Southern California, Los Angeles, California and 6900 W Interstate 40, Suite 100-03, Amarillo, Texas.
4. Narula, Ramesh K., "Profitable Growth Through QS/ISO 9000 — A Case Study in Breakthrough Thinking," Third International Breakthrough Thinking Gathering, Los Angeles, CA, June 22, 1996.

Chapter 8

Concept Management: Working Together

*Seeing isn't believing,
believing is seeing!*

She was walking along the path feeling rather depressed and downtrodden. Her chin lay heavily on her chest and her eyes seemed fascinated with the path in a dazed, dumbfounded sort of way. She was having a rough day, and the last thing she wanted was to talk to someone.

A small boy was coming up the path toward her in the opposite direction. The boy bounced and hopped along carrying a stem of beautiful white orchids that he had picked a few minutes earlier. He was getting a little tired of carrying the flowers — it restricted activities — and when he saw the lady up ahead he thought he would give them to her. Maybe it would cheer her up.

He bounced, hopped, and skipped his way up to the lady and without saying a word held the flower out to her. The lady was astonished, but accepted the unexpected present. She wondered and watched in amazement, and the boy continued on down the path. Then she looked at the orchid. It was the most beautiful flower that she had ever seen. And the impulsiveness of the gift seemed to make it even more beautiful. She now had a mission. She would hurry home and put the flower in a vase with water. She knew that if the flower didn't get put into the water soon, it would start to dry out. She hurried on home. A flower this beautiful had to be preserved as long as possible.

When she arrived home she went to the cupboard and got out her vase. It was dusty, far too dusty to hold a flower this beautiful. So she proceeded to wash the vase carefully. Then she filled it with water and put the flower in. As she approached the mantle where she knew a flower this beautiful belonged, she was disappointed by how dirty and dusty the mantle looked. So she proceeded to wash and dust the mantle. Then she got out her best doily, placed it on the mantle, and put up the vase with the beautiful orchid. She stepped back to admire the flower and was disgusted with how dirty the surrounding wall was. The wall took away from the beauty of the flower. She proceeded to clean that wall, and then all the walls, and the floor, and before long she found herself cleaning everything in the house.

The orchid had given the lady a mission and a purpose. She was excited about cleaning the house and found it to be fun, now that she had a reason to do it. And by having a clean house she found herself feeling better about herself.

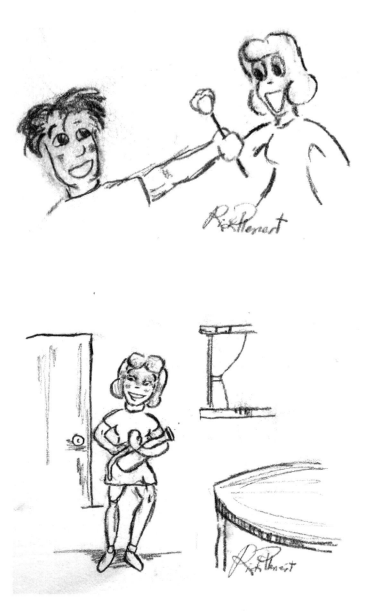

Is this a silly story, or is this a depth-filled example of how a small act can make an enormous difference? When was the last time we engaged in a "random and selfless act of kindness?" Don't we wish someone would do it for us sometime? We spend our lives judging others by their actions, but we want others to judge us by our intentions. If our intentions are not converted to actions, then what good are they?

> *We don't want God to remember our sins, so there is something fundamentally wrong in our relentlessly trying to remember those of others.*
>
> **Elder Jeffrey R. Holland**

This chapter will refocus on people interaction, which is extremely critical to the success of Concept Management effectiveness. The principles of people interaction (working together) that need to be stressed are

- Teaming
- Creative synergy
- World class person
- Globlocal[1]
- Conflict reduction
- Believing and doubting games

Teaming

> Teaming is like flossing, everyone says they do it but nobody really does!

The topic of teaming has been beaten to death, both in this book and in numerous outside sources. We can go all the way back to quality circles, or move forward to modern-day team building techniques. Unfortunately, we still don't have it right. What we engage in is grouping (throwing people together for quick fixes) rather than teaming, which focuses on long-term synergistic improvements through long-term, carefully developed, interactive relationships. We're not going to repeat a discussion on the benefits of teaming here. Return to Chapter 3 to learn about World Class teaming, or review the chapter on teaming and Total Quality Management in *World Class Manager* for more information.[2]

There is one element of teaming that is, however, somewhat innovative and falls into the creativity elements of Concept Management: the quality council. Under Concept Management philosophy, if team synergy is good at the bottom or in the middle, then it should also be extremely valuable at the top. The quality council, as defined by Concept Management, would be a form of team-based CEO. There is still a CEO, but this is an interface individual who works between the council and the various stakeholders, such as the board of directors, the owners, major customers, governments, and special interest groups. The council runs the company. This is a move away from the traditional "king of the hill" approach to management to a more democratic form. If democracy is good for governments, why shouldn't it be good for corporate management as well?

Creative Synergy

> *The best way to have a good idea is to have many of them.*
> **Linus Pauley**

Creativity can occur in the individual. However, when individual creativity is shared with others, through the perspective of others, synergy occurs. The first person comes up with an idea, and the second person expands, enhances, and improves on the idea because of his or her unique perspective. The second person probably would not have come up with the idea, and the first person probably would not have come up with the expansion on the ideas because of the limitations of only one perspective.

Intel moved a research center to Malaysia because it recognized the dynamics of synergy. In Malaysia, the population is 60% Malay Moslem, 30% Chinese Buddhist, and 10% India Hindus. Mixed into this pot are some Christian (United States and European management) and Shinto (Japanese management) perspectives. The dynamics of having a broad mixture of cultures and perspectives has generated ideas that would never have come out of only one individual, or even out of just one culture, such as a U.S. Western Christian culture. In Concept Management we need to stress this synergy and focus all things, including the measurement system, on motivating interaction and synergy.

World Class Person

> *The highest reward for a man's toil is not what he gets for it but what he becomes by it.*
> **John Ruskin**

Concept Management places a specific set of requirements on individuals. You cannot manage an organization or be a team player in an organization that focuses on waste elimination, the elimination of nontrust systems, teaming, cooperation, etc., without corporate values, and without individual values. Integrity is not just a buzzword or an ideal, it is a requirement. You need to be able to trust your employees, and they need to be able to trust you.

Integrity is saying what you do and doing what you say!
Gerhard Plenert, World Class Manager

World Class Person characteristics include the following:

- Integrity
    Ethical and honest
- Standards and a value system
    Goal-focused in your personal life
    Loving, humble, moral, and obedient
- Society value-adding
    Job enrichment oriented
- A Leader
    Exemplary
    Enthusiastic
    Compromising
    Understanding
    Self-replacing
    Environmentally conscious
    Worker-safety conscious
    Open, sharing, and fun-loving

These areas are all discussed in detail in the book *World Class Manager*.

A World Class person is one who loves learning, and loves to see others learn. It is an individual who likes change and sees opportunity, rather than difficulties, in the change process. To become a World Class organization you first need to become World Class people.

The towers of tomorrow are built upon the foundations of today.
Alfred Lord Tennyson

Globlocal

> We make our own friends, but God made our neighbors; then he commanded us to love our neighbors as ourselves!

Globlocal, simply stated means *plan globally but act locally*. The world is a big place, with lots of perspectives. No perspective is so correct so as to make all other perspectives incorrect. There are simply degrees of correctness, and there are situations where one perspective is more correct than another. The U.S. perspective, simply because of the country's size and level of technology, is not more correct than everyone else's. Might does not make Right! For example, no religion is totally wrong. There are degrees of value that can be gained from all religions, and to treat any religion with disdain, without studying its potential value first, is childish. Is it offensive when an atheist (a religion that believes in the nonexistence of a God) puts more faith in science (a belief system that spends all its time trying to prove itself wrong)? Religion and science both have value, and they are both searching for truth. And, since there is ultimately one truth, all seekers of it should eventually end up at the same place. What is disturbing are those individuals who reject the feasibility of any truth-seeking perspective outside their own, regardless of what label we place on it.

Globlocal is the recognition that we are not international, but inter-people. Globlocal is a search for the uniqueness, realizing that there is synergy in that uniqueness. If we are rubber-stamped out of the same mold, we are just a copy. The American perspective on any issue, such as ethics, trade, competitiveness, quality, productivity, value-adding, etc., is a minority perspective. The world is filled with many other perspectives, and the more we listen the more we learn.

> The best listener is the one with their mind open and their mouth closed!

Conflict Reduction

A man is about as happy as he makes up his mind to be.

Abraham Lincoln

In the area of people interaction (working together), conflict reduction becomes a major issue, and one of the primary purposes of this chapter is to help us recognize how conflict should be handled, and what the Concept Management approach to conflict reduction should be. Let's start with a very simple example: husband–wife conflict. If a couple gets into serious disagreement, the traditional approach toward resolution is to focus on "root cause" analysis. We try to blame the problems on an inadequate childhood, or on abusive parents, or on improper school, or sexual inadequacies, or being too busy at work, etc. We hope to find some root cause on which we can blame the conflicts. Often, this root cause analysis is performed with the help of some social worker or psychologist, who has considered him- or herself an expert at identifying root causes. Then, with this root cause in hand, we have an excuse to file for divorce, because "we can't live with that situation" or because the "spouse was deceptive in not supplying all the information about his or her past." We have become a culture that no longer sees an inappropriate family life, or sees inappropriate lifestyles as sins; we see them as syndromes. Perhaps a better word for syndromes is *excuses*.

The next step, after we have accomplished our root cause identification of the problems, is to run to the lawyers. And we know that lawyers are not interested in resolving conflict, since they earn their incomes by (increased) conflict. As a result, the relationship becomes even worse, now focusing on anger, fighting, and, in the end, divorce.

The Concept Management approach to conflict reduction is to ignore root causes. Rather, the focus should be *re*focused on purpose expansion. For example, the conflict should be handled through a series of questions such as:

    What is the purpose of the fight?
        To have a better life.
        To establish common ground.
        To win.
    What is the purpose of having a better life?
        To achieve our life's goals.
        To have harmony in the family.
        To help our children achieve greatness.
        To get closer to God.

The purpose expansion would continue on in this manner and, rather than having animosity continue to develop between the husband and wife, they would soon come to realize that they really both want the same things. They would find their common ground. They would realize that the fight was actually keeping them from solving the differences. The fight would suddenly become silly, and compromises could now be discussed in light of the larger, ideal purpose of their relationship.

This approach for conflict reduction should be used in any disagreement. Once commonality is achieved, compromise can begin, and solutions can be found. The conflict is usually a hindrance to successful solutions, rather than an aid. This approach is similar to the Chinese box discussed earlier. For example, in the diagram below we can see that we are fighting (conflict) at one level, but that solutions are found at a completely different level, and that conflicts seldom bring about the win–win situation that Steven R. Covey stresses for highly effective people to have.[3]

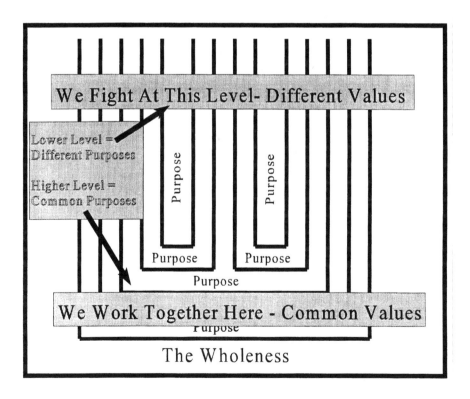

Showa Ceramic Company was engaged in a difficult and challenging management–labor conflict. Management, which was trained in concept creation techniques, convinced the union leaders to sit down with them and do a purpose expansion of the situation. The two sides discussed why the salary conflict existed, they discussed the poor economic conditions, and how an increase in sales would increase the corporate economic situation. They realized that they were both focused on a common purpose, stability and growth. They realized that analysis, conflict, and criticism would only lead to bankruptcy, and that neither side would achieve its objectives in this situation. They came up with a compromise agreement that tied salary to the company's economic position; if economics improved, so did the salaries. They learned that integration and a focus on purpose expansion was the key to conflict resolution.

For additional reading, let me suggest William Collison's book on conflict reduction.[4] There is no magical solution to all conflict; however, the Concept Management approach has been extremely successful.

Believing and Doubting Games

In Concept Management, the teaming process utilizes an approach to minimize conflict and to maximize synergy. The dynamics of teaming requires that we spend a large amount of time developing relationships. For example, in the Intel example in Malaysia that was discussed earlier where we had a mixture of many cultures, we would often find that the most prestigious person, or the individual with the most seniority, or the individual with the most education, would dominate the discussion process. All individuals in the team would tend to subordinate their opinions to the individual with the seniority. Similar conflicts occur in all teaming situations.

In order to resolve the superiority conflicts that occurred, we needed to focus on teaching individuals that they criticize systems, not individuals; that there was no need for personality attacks. It is the system that constantly needs change and improvement, not the individual. Systems are inferior and need correction.

Another technique that is helpful in developing team synergy is the believing–doubting game.[5] In this approach, we play a game, and playing games tends not to be as confrontational. Here is how it works:

Believing Game
✤ (plus) — adding ideas
✖ (times) — team synergy
Focus — to build trust and confidence
To encourage imagination

Approach — vague ideas
　　　　Focus on the whole
　　　　Nonconfrontational
　　Procedure
　　　　Purpose expansion
　　　　Focus purpose definition
　　Generate lots of options
　　　　Just keep throwing out lots of ideas
　　　　Nothing is criticized, everything is accepted
　　Doubting Game
　　　　− (minus) — criticize, find what's wrong
　　　　÷ (divide) — analyze
　　Focus — idea evaluation
　　Approach — logic
　　　　A search for clear approaches
　　　　Identify the part that is useful, implementable, usable, effective
　　Procedure — challenge the workability of the ideas

In the believing–doubting game, the objective is to first generate as many ideas as possible. We want to explore all the options. The doubting process then focuses on idea selection. We are searching for the highest level idea that is workable and achievable. The only precaution is that you spend a sufficient amount of time on the believing process. Don't rush through this process; this is the creativity process. Avoid the tendency to start the doubting (analysis) process. That's not the true objective of this game. If this is a problem area for you, call separate meetings for each process. The first meeting is a believing-only meeting for idea generation purposes. The second meeting starts off as a believing meeting, because often new ideas are thought of between the meetings. Then the doubting game begins. From this process we then get a selected alternative.

Since games tend to be less confrontational, the believing–doubting exercise makes a game out of idea generation. It also removes some of the situational and environmental limitations that are often placed on creativity.

Summary

Happiness does not depend on what happens outside of you, but what happens inside of you. It is measured by the spirit with which you meet the problems of life.

President Spencer W. Kimball

This chapter refocused us on people interaction. People can make any project a success or a failure, including Concept Management. With people ownership, the program is a guaranteed success; however, without it, it is a guaranteed failure. Therefore, this chapter offers a few tools to assist in the development of the proper kinds of people interaction, such as

- Teaming
- Creative synergy
- World class person
- Conflict reduction
- Believing–doubting games

These tools are just stepping stones. They can't replace the commitment and involvement of all levels of management, including the quality council. But they can help you get a little bit closer to Concept Management success.

> *... it is not enough just to be good. You must be good for something. You must contribute good to the world. The world must be a better place for your presence, and the good that is in you must be spread to others.*
>
> **President Gordon B. Hinkley**

Endnotes

1. Sometimes this is referred to as "Glocal." We have chosen the term "Globlocal" in this book. Both terms refer to a global–local perspective.
2. Plenert, Gerhard, *World Class Manager*, Prima Publishing, Rocklin, CA, 1995, pp. 261–279.
3. Covey, Stephen R., *The Seven Habits of Highly Effective People*, Simon & Schuster, New York, 1989.
4. Collison, William, *Conflict Reduction: Turning Conflict to Cooperation*, Kandall/Hunt Publishing, Dubuque, IA, 1988.
5. Elbow, Peter, *Writing Without Teachers,* Oxford University Press, New York, 1973.

Chapter 9

Concept Management: The Road to Success

> *First say to yourself what you would be;*
> *and then do what you have to do.*
>
> **Epictetus**

The worker was fearful; he had a desperate decision to make. The oil platform he was standing on was aflame. He could stay on the platform and try to ride out the fire, but he would probably be badly burned, or he could jump into the cold, rough waters of the ocean. The decision process was simple:

- Get burned — it's easy but the results are questionable.
- Jump off — it's risky, you may freeze or get hurt, but your chances of long-term survival are much greater.

Isn't that what Concept Management is all about? We could stay on the burning platform (traditional methods of analysis and change) and continue to fight the little fires, hoping that by some wild chance we may survive and not get burned. Or we could take the plunge and dive into the scary ocean of Concept Management, looking for that edge that will give us long-term survivability.

In the 1970s, Toyota Central Research Labs, a private, profit-oriented corporation separate from Toyota Motors with about 1,000 employees, had a clearly defined sequence of steps that it utilized to push products

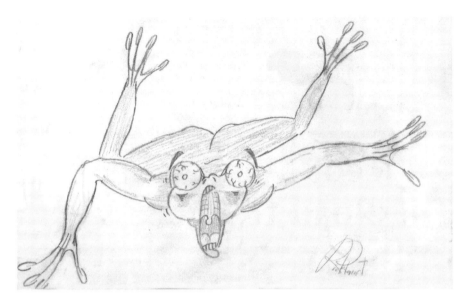

to its customers. It used the research approach, which brought with it the attitude that the corporation was smarter than its customers. It would develop what the customer needed, and then push the product out to the customer. This pipeline push form of research is now considered narrow thinking, since it is not related to customer or human needs. The Research Labs was on a search for truth, not solutions, and the success of the organization was deteriorating quickly.

In the 1980s Toyota Central Research Labs realized that it needed to change its approach or the labs would soon be closed. The leadership started Concept Management training, focusing first on Concept Engineering (Breakthrough Thinking). They studied purposes, values, and measures, and developed a new theme for finding solutions. They introduced a corporate-wide management philosophy change, developing new values. They stressed that research would now be centered on finding solutions, not on finding truth. Thus, they switched to a *pull* approach, where the customer pulled the needed research. It became a customer- and people-driven research organization.

A counterpart to Toyota Central Research Labs, Toyota Motors, utilized the Concept Management system's view to analyze its production process. It focused on workplace design and developed a process called Shikumi Zukuri. Shikumi (system) Zukuri (create or design) focused on the development of a interrelated, integrated concept of work-flow and work-design that is now called JIT (just-in-time). "System" does not mean "computer system," it means the organization of parts to create an integrated whole.

Concept Management: The Road to Success

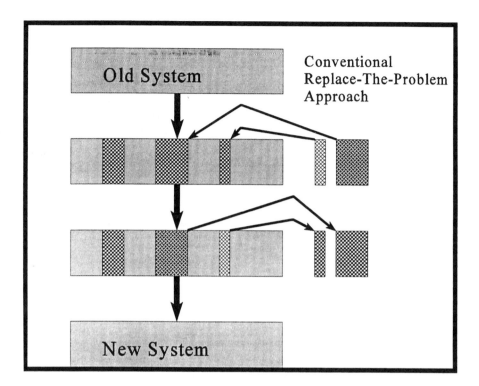

For Toyota Motors, this integration encouraged their 1960s development of quality circles, JIT, Kanbans, and what is now referred to as the "Toyota Production System," a creative production systems model that has defined a new world standard for productive effectiveness. Companies the world over have attempted to copy pieces of the Toyota concept, but they continually miss the big picture, the Concept Management integrative focused picture. It is the big picture, not the small pieces, that has made Toyota so successful.

Concept Management is a break, freeing ourselves from traditional, Descartes-style replace-the-problem thinking. In the diagram we see how this type of thinking works. We are simply putting out fires as quickly as possible, only to discover that another fire has reared its ugly head. We are not looking at the larger perspective. In the diagram we see how traditional root-cause analytical methods have us searching for plug-in solutions. We replace small pieces, never asking if the larger piece should even exist.[1]

With Concept Management we are no longer moles, digging holes in the ground — as we analyze, we dig deeper and deeper holes, and soon find ourselves buried and going nowhere. With CM we are now searching for a higher meaning, a relevance to all things. As in Zen, the purpose

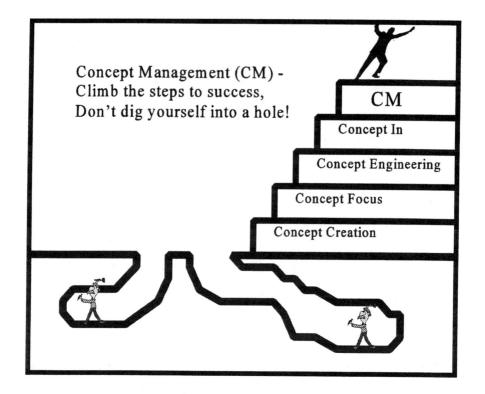

of life is to find that higher meaning, not to find the root causes. Concept Management is taking the stepping stones to organizational higher meaning, as shown in the diagram above.

Reviewing the principles of Concept Management, we have

- Concept Creation
- Concept Focus
- Concept Engineering
- Concept In
- Concept Management

Concept Creation is taking a step back and asking "What is the purpose of ...". Through this process of purpose expansion we open up a whole new world of ideas, options, and opportunities.

Concept Focus is the taking of that higher purpose and building a goal structure around that purpose. It is the definition of values, core competencies, visions, missions, and strategies. And it is the development of measurement and motivation systems that stimulate the proper responses around those strategies. Concept Focus is also the "people" element of working together, focusing on teaming, gainsharing, and empowerment.

Concept Engineering is the taking of creative ideas and using the focus to develop consumer-oriented, people-friendly products through a pull strategy, similar to Toyota Central Research Labs as related at the start of this chapter.

Concept In is the development of markets for the created and engineered ideas. It is getting the products to the customers that had originally defined the needs to begin with.

Concept Management is the integration of these process with a systematized perspective. With Concept Management, we eliminate the negative effects of analytical thinking, such as reduced creativity and increased contention through defensiveness. Instead, we push the organization to a higher plain, and then bring it back into reality. The chart below compares the long-term perspective and effect of the idea creation process.

Under conventional thinking, we move backward (downward). In the diagram below, Area A is for problem identification and the development of an action plan.[2] Areas A and B identify the analysis process that occurs. Here is where we go through extensive data collection in an attempt to identify the root causes of the problem under study. Creativity actually goes down as we do root-cause analysis, because we become narrow and

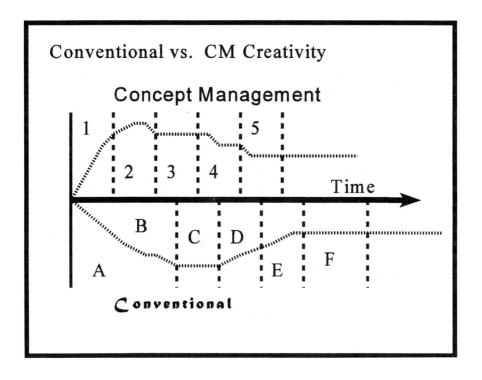

focused. Area C is where we take the data and develop an idea for solutions. Area D is where the target solution is defined. Area E is the engineering and development of the idea, and Area F is the implementation, feedback, and control process.

With Concept Management, Area 1 is an area of increased creativity through purpose expansion and idea generation. Area 1 is Concept Creation. Area 2 is Concept Focus, where we define the values, goals, and measures of the concept created. Area 3 is Concept Engineering, where the idea is developed, and Area 4 is Concept In, where the marketing of the idea occurs. Area 5 is Concept Management, the overall, ongoing management of the concept created.

Nippon Electric Corporation (NEC) uses Concept Creation (Breakthrough Thinking) and Concept Focus (World Class Management) principles for the production of its semiconductors. Its semiconductor division of 3,000 employees has generated a $400 million U.S. per year reduction in production costs on their production line. They received the "President's Award" for the innovativeness and creativity that eventually resulted in their redesign of the entire production line.

The NEC approach stressed "learning from the future." It focused on visioning and the development of holonic products. Its focus was to look at society, then at the business domain, then at the products that this business domain should focus on (see diagram below).

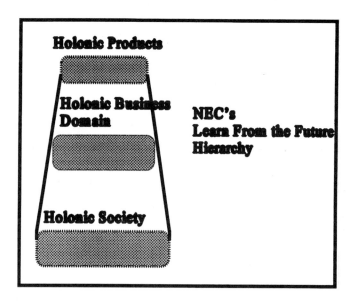

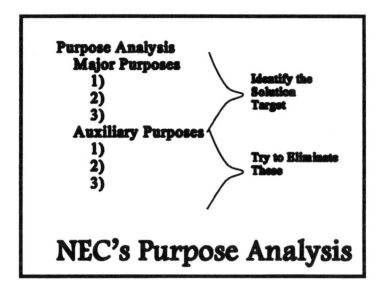

NEC identified a solution target, then they developed a road map of how to arrive at the target by exploring a purpose expansion (see diagram above). Through this expansion they identified major purposes and auxiliary purposes. The major purposes became the target objectives, and the auxiliary purposes were considered to be waste, because they distract from the target or focus purpose.

Through this process, NEC achieved approximately a 50% cost reduction in its production line. Additionally, it applied these techniques elsewhere and achieved a total cost reduction in all factories throughout the organization. The process has been so successful that NEC has established an internal training program focused on making everyone an expert at Concept Management principles.

Aishim Precision Equipment manufactures gears, transmissions, and brakes for Toyota. During the 1960s it utilized Concept Creation and Concept Engineering principles to redesign its factories. It utilized "visioning," and continues to use it today as it looks forward to the 21st Century. Each department created its own targets (vision, mission, ...). They developed them separately, but kept them integrated into the corporate vision. Aishim is now aggressively preparing to compete throughout the next century.

Summary

Concept Management is people, philosophy, and it is principles. It isn't clearly defined, specifying which procedure to take and when; however it does have guidelines for success. It specifies the use of Breakthrough Thinking (BT) principles in Concept Creation, World Class Management (WCM) principles in Concept Focus, and both of these along with TQM in Concept Engineering. It specifies the need for Concept In marketing principles to get new, visionary products out to the customers. Concept Management stresses the importance of people interaction, ownership, motivation, and participation. It focuses on a systems approach and futuristic "pull" mechanisms that will get us there before the competition.

Concept Management principles have been the success story in Japan for a long time. The United States has attempted to copy parts of the process, but repeatedly keeps missing the larger message of integration, change, and a forward-looking perspective. Concept Management is the future.

There is a saying: You can count the number of seeds in an apple (conventional analytical thinking), but you can't count the number of apples in a seed (futuristic thinking). We need to quit focusing on how many apples we have today, and start focusing on the potential of an eternal apple harvest.

Endnotes

1. Nadler, Gerald, *The Planning and Design Approach,* John Wiley & Sons, New York, 1981.
2. Nadler, Gerald, *The Planning and Design Approach,* John Wiley & Sons, New York, 1981. This diagram is also part of the Center for Breakthrough Thinking, Inc.'s model for the "Potential for Ideal Results."

A NEW WORLD ORDER OF CHANGE

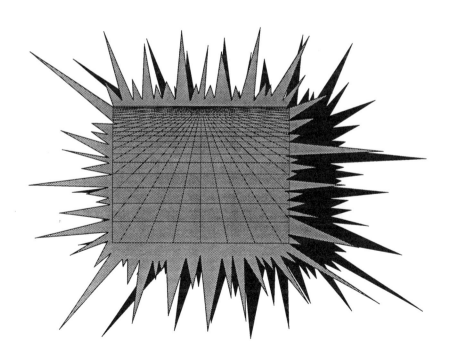

Chapter 10

Concept Management: The Test

An employee eagerly delivered a report to Henry Kissinger. Kissinger held on to the report for a couple days, then called the man into his office and asked, holding out the report, "Is this the best you can do?" The man took the report, reworked it, and brought it back to Kissinger. Kissinger again held on to it for a couple days and then called the man into his office once again. Again he asked, "Is this the best you can do?" The man sighed with frustration, took the report, and once again reworked it. Kissinger repeated the process a third time, asking the individual if this was his best. The man threw up his hands and told Mr. Kissinger how much effort he had put into the report and how he couldn't think of any way to improve the report any further, to which Henry Kissinger replied, "Good, now I'll read it!" And this time he did.

Are you a Concept Manager in your thinking? We would all like to think that we are innovative and leading-edge individuals, but sometimes we need to be brought back down to reality. That is the purpose of this short test. It is a reality check to see if you are on track.

Question Set Number 1 — Did you find the Henry Kissinger story offensive? Were you irritated by the fact that he didn't look at the earlier reports? Looking at the earlier reports would have been a waste of Kissinger's time reading it, and a waste of the author's time writing it. Do we accomplish work "just to get it over with?" If that's the case, then why do we waste our time doing it? We would be better off doing one job well than wasting time on five jobs poorly done. Reports don't have to be long to be good, they have to be meaningful.

Question Set Number 2 — You hear that in your factory the production flow line for valves is starting to have a high breakdown rate and that it will soon need to be replaced. You put one person to work estimating the costs of the breakdowns, the production time lost, and their frequency, and put someone else to work getting quotes on the equipment necessary for a new line so that in the next production meeting we know what our options are. WRONG! You should first ask, "What is the purpose of fixing the production line?" and then continue on with a purpose expansion of the entire process. You'll be surprised how this will lead you to alternatives you never even considered.

Question Set Number 3 — You've given an employee an assignment to reorganize his or her work area. You have some idea of how it should be done, but you decide to leave the employee alone to do it on his or her own. However, what the employee comes up with is so unworkable that you decide to intervene and help the employee "see the light." WRONG! Give the employee a chance. You might be pleasantly surprised to find out that it's really you who hasn't seen the light.

Question Set Number 4 — You asked your employees to come up with a gainsharing program that would properly motivate them to work on corporate goals. The employees come up with a program that would require them to have access to confidential accounting reports, so you reject their plan and ask them to try again. WRONG! You've just destroyed their confidence in you and destroyed all your credibility. What they'll do next is try to figure out what it is you really want, and then just regurgitate it back to you. Innovativeness and creativity have just been destroyed. If you really didn't want them to do it, if you had some pat answer that you wanted them to come up with, then you shouldn't have asked for their opinion.

Question Set Number 5 — You are asked to predict next year's sales of a new product line. You get the computer techies to generate forecast projections for you based on the performance of a similar product line. Then you utilize these numbers to create your projections. WRONG! The past is no better at predicting your future than a 5-year-old child is at predicting if his or her marriage will be a success.

Question Set Number 6 — You've just discovered that one of your employees has swindled the company out of $5,000 by manipulating the accounts receivable program. The employee found a loophole that allowed him or her to transfer rounding errors into his or her own payroll account. You decide that what you really need is to get rid of the antiquated system that you have and to get a new accounts receivable package that wouldn't allow this situation to happen again. WRONG! What you're really saying is, "That was too easy. Let's give the employees a new challenge and see if they can break through that one." My family has a dog in the backyard and we have an area fenced off for him. He takes it as his personal challenge to see if he can escape this fenced-off area. My two sons (ages 16 and 14), myself, and my wife have taken it upon ourselves to try to keep the dog in the fenced-off area. We have improvised, innovated, changed, and fixed fence for months, but after a couple of weeks the dog will have found a new way to escape. Sometimes I have to wonder who's the master of the fence. I think that he simply sees the fence as a challenge, and he's going to win. Employees are that way too. Rules, regulations, and limitations are challenges. Get rid of the rules, and you get rid of the challenges. Trust the employees, and they will trust you. Demonstrate integrity, and you will receive integrity. Often the barriers we build are in areas where we trust ourselves the least.

The faults we see as worst in others are often the same ones we try to hide in ourselves.

Question Set Number 7 — When was the last time you got rid of a system? Or, are you like the government, which only knows how to add committees but never knows how to get rid of any? This is the bureaucracy growth that has become the death of many large companies. Do a purpose evaluation of all your "systems," including the computer systems, the accounting systems, the personnel systems, etc. Then let the employees be a part of this purpose expansion process. Any system that doesn't have a purpose should immediately be eliminated.

Question Set Number 8 — Draw a time clock of all the functions that you performed during the last 24 hours. List all the tasks that you accomplished. Now ask yourself how much of the last 24 hours was spent focused on the goals of the company. How much of that time was value-adding? How much of it was a waste? Would any customer know if you were to suddenly disappear? Would it make a difference to the success of the company? If not, why not?

Question Set Number 9 — You have a problem and so, in your innovative way, you attack that problem head on, using everything you know. When was the last time you asked an employee, specifically a subordinate, to share their opinions or ideas about how the problem should be handled? Or do you consider yourself superior and wiser than those "flunkies?"

Question Set Number 10 — Do you hold a meeting because you like to hear yourself talk? Or is there a purpose for the meeting? Never hold a meeting unless you have a defined purpose; otherwise, it is a waste. Remember, *meetings don't make teams, and teams don't live by meetings.*

Question Set Number 11 — Is the performance of your employees measured so that their success will make you look good in the areas in which you are measured? Are you measured so as to make your boss look good? If the measurement system is not properly focused, the organizational goals have little chance of being achieved.

Question Set Number 12 — Can you recite the vision and mission statement of your organization? Do you even know if it has one? Are your efforts focused on achieving this vision, or do you focus on "keeping your job and getting a raise?"

Question Set Number 13 — Can your employees freely walk over and talk to an accounting clerk, a marketing manager, or the CEO, or are there certain people that they shouldn't talk to without seeing you first?

Question Set Number 14 — Have you had your employees write a "blind" performance evaluation of how you perform as a manager? Focus this on the principles of Chapter 3 and Chapter 8? Does the prospect of this type of evaluation scare you? Should you do an evaluation of this type for the managers who work for you?

Question Set Number 15 — How easy would it be for you to change, or better yet, throw out the costing system in your company? Are you committed to the system you have because "it has always been done that way" or because the "tax man wants it that way?"

If you've answered these questions with answers that demonstrate your forward thinking and World Class Management abilities, then you are ready to progress along the path of Concept Management. A key to Concept Management success is the understanding of all the principles discussed in this book, including

Breakthrough Thinking
World Class Manager
Total Quality Management

Study these principles as discussed in earlier chapters and in the recommended readings. Then go on to focus on the Concept Management principles discussed in Chapter 5. Follow the steps outlined in Chapter 6 toward becoming a Concept Management organization, and then incorporate the enhancements discussed in Chapters 7, 8, and 9, and you're on your way!

Concept Management: The Test

Summary

The Micronesian sailor lies on the bottom of his canoe, listening and feeling the motion of the waves. From this motion he can discern a message — a message of which islands the waves have encountered, and how long ago, and which other waves these waves have encountered. From this message, and using his training, experience, and knowledge about the waves, the sailor knows how to determine what message they are trying to tell him. He knows where he is and how to get where he wants to go. From the message of the waves, he knows which way to find home.

More recently, technology has removed the need to understand the message of the waves. Perhaps we need to once again listen to the waves and rediscover some of the lost principles that have made us a great nation, principles that focus on "We the people" and the synergy of mixed cultures and peoples. We need to refocus on ideals like God and country. We need integrity and trust, trust in each other, in our government, and in our business leaders. But trust must be earned, both by us and by our leaders.

Concept Management is an exploration of the new, but that exploration is built upon strong principles and values of old. We need principles and values to put us on the right track toward our future.

> *Practically all dishonesty owes its existence and growth*
> *to this inward distortion we call self-justification.*
> *It is the first, the worst, and most insidious and*
> *damaging form of cheating — to cheat oneself.*
>
> **Spencer W. Kimball**

Chapter 11

Concept Management in Your Future

> *... without patience, we will learn less in life.*
> *We will see less. We will feel less. We will hear less.*
> *Ironically, "rush" and "more" usually mean "less."*
>
> **Neal A. Maxwell**

She watched with excitement as they filmed Richard Dreyfuss at John F. Kennedy High School. This was her school, and the story of budget cutbacks and of the elimination of the school music program was close to her heart. She anxiously anticipated the release of the new movie, *Mr. Holland's Opus*, that was being made. However, she had terminal cancer and it was questionable if she would live long enough for the movie to be released. On some days it seemed like she could hold out forever. On other days it seemed like just one more day would be too much to suffer through.

One particular weekend, after a visit to the emergency room, the doctors predicted that she would only have a few days left, and they sent her home to spend those last few days with her family. Her husband, on a fluke of inspiration, wrote a letter to the Disney Corporation and told them about her predicament, and stressed how much seeing the movie would mean to her. The Disney people called immediately to the doctors to verify the health risk, and then sent an employee who would hand-carry a copy of the yet-unreleased film to Portland.

Her husband lovingly helped her from the bed to which she had been confined, and from which she could barely move. The family and the Disney employee sat together to watch a private and very emotional screening of *Mr Holland's Opus* in the living room, where no one held back the tears. After many thank-yous, the Disney employee left and returned home to Los Angeles. Two days later she passed away.

To say that this viewing was the highlight of her life would be going too far. But it could easily be said that when corporate America takes the time to show compassion and love for an unknown individual, then perhaps there is still hope for us all.

> *Men honor what lies within the sphere of their knowledge, but do not realize how dependent they are on what lies beyond it.*
>
> **Chuang-tse**
> **(as quoted in *The Te of Piglet* by Benjamin Hoff, p. 133)**

Concept Management is fun. It is so exciting to go into a company and to make a difference without being confrontational. It is exciting when the employees realize that change does not threaten their existence under the disguise of downsizing or cost-cutting. It is exciting to watch the eyes and faces of employees light up as they expand their thinking to new ideas, beyond what they had previously considered. It is fun to watch employees innovatively come up with ways to make improvements, not just at work but also in their own lives.

Employees latch on to Concept Management and take ownership in its philosophies. Here is a tool that they can utilize to make a difference, to give themselves focus, and to give the entire organization direction. This book was written to assist the manager, but the same principles are eternally true and useful in changing personal lives, families, career paths, etc. These principles are exciting, and we invite you to reread this book along with *World Class Manager, Breakthrough Thinking,* and *Creative Solution Finding,* to broaden your perspective to an entirely new world of innovation. Let's all become World Class Concept Managers together.

> *One's ideas must be as broad as nature if they are to interpret nature.*
>
> **Sherlock Holmes by Sir Arthur Conan Doyle**
> **(as quoted in *The Te of Piglet* by Benjamin Hoff, p. 132)**

We invite your letters and are excited to hear from you. We want your stories, successes, frustrations, and ideas. Concept Management is not a wall, it is an organism, constantly changing, and constantly developing. Please share your experiences with us at the addresses given in "About the Authors" following the Table of Contents in the beginning of this book.

Let's end this book with a story about the importance of a paradigm shift in our thinking:

She was frustrated. She felt that her life had become a soap opera of sorts, filled with struggles, challenges, and frustrations. The bill collectors were constantly hounding her. She wasn't sure where the next dollar was coming from. Today had been an exceptionally difficult and tiresome day, and she was finally going to have a chance to sit down and watch her favorite soap. What she didn't want to do was answer the door. Some persistent bill collector was repeatedly knocking and she just couldn't face him. But the persistence of the visitor had won out. She couldn't put up with the knocking any longer, and she went to answer the door.

"Good Morning," was the pleasant response she received for opening the door.

"Hi," she responded in a leery, mistrusting, cautious manner.

"I've come to offer you a gift."

The lady was relieved that it wasn't a bill collector. Her mind was already developing a strategy for getting rid of this salesman.

The salesman held out a small black box in the middle of which was a big red button. "This is for you," he said. "You can have this box. And all you have to do in order to receive one million dollars is to push the red button. However, when you push the button, 'someone who you do not know and who does not know you will die'."

The lady was astonished. Of course she would like the million dollars. But to have someone else die for it was incredible. She would never push the button. She was about to slam the door but the insistent salesman had already wedged his foot in the doorway. She could see that he wasn't going to leave until she accepted the box. In the end, she decided to take the box, just to get him to leave.

Back alone in the house, she looked at the curious box. It showed no signs of technology. It was simply a small black box with a red button. She decided to put the box somewhere in the back of her cupboard, out of reach and out of mind. But the box wouldn't leave her mind. At first she was bothered by the audacity of it. She was given the power to end someone's life. She was sure she could never push that button.

But then she started to become bothered by the money. A million dollars would sure solve a lot of her problems. And the more the bill collectors hounded her, the better the million dollars sounded. Soon her mind was telling her, "What's the difference. The person that would die was probably someone old or sickly anyway. They might be better off dead. I could probably do more with the million dollars then they are accomplishing with their lives."

As the days went by it became easier and easier to consider pushing the button. Soon the million dollars became more important than the life of "someone she did not know and who did not know her." And finally, one day, an exceptionally tiresome day, in a fit of frustration, she found the black box and, holding it in her hands, stared at it. The realization of what she was about to do struck her and she was at first bothered, but slowly her mind brought her back to all the old arguments of how it really didn't matter and how important the million dollars were, and in a surge of anxiety, she pushed the red button.

Almost immediately the doorbell rang. She peeked out the window and found the salesman that had originally delivered the black box standing there. How did he find out so quickly that she had pushed the button? Oh well, he was here to give her the million dollars. She swung the door open and there he stood, with the same big smile that he had before. He held out a check to her and said, "Here's your one million dollars. It's all yours as soon as you hand me the black box."

She was surprised that he wanted the black box back, but her eagerness to receive the million dollars made this a small concern. She handed the box back and eagerly took the check.

The salesman started to turn and walk away, and almost as an afterthought she asked him, "What are you going to do with that box?"

The salesman smiled at her and said, "I will give it to someone whom you do not know and who does not know you."[1]

Concept Management breaks us free from traditional paradigms and allows us to focus on the future. Any company that tries to be futuristic without Concept Management will find itself bogged down with traditional analytical thinking and paradigms. It will analyze and justify its situation until it finds itself pushing the button on the box (like downsizing), only to find out afterward that the button will soon be pushed on them (like bankruptcy). With Concept Management, an entirely new world of opportunities opens up. New ideas and new synergies are created. Being forward-looking takes on an entirely new meaning. Methods, techniques, and the very role of employees changes. Competitiveness gets redefined. Paradigm shifts in thinking take place. An entirely new world of opportunities opens up. Let's all move forward to become World Class Concept Managers.

When one tries to rise above nature, one is liable to fall below it.

**Sherlock Holmes by Sir Arthur Conan Doyle
(as quoted in *The Te of Piglet* by Benjamin Hoff, pg. 132)**

Endnotes

1. This story was adapted from *The Twilight Zone*. It was also told in Plenert, Gerhard, *World Class Manager*, Prima Publishing, Rocklin, CA, 1995, p. 279; Adolphson, Don, Matthew DeVries, and Heikki Rinne, "Rethinking Business: A Broader Sense of Responsibility," *Exchange*, Fall 1994, p. 5–9.

Index

A

ABC analysis, 60
Aishim Precision Equipment, 149
American Civil Liberties Union (ACLU), 119
APICS, vii
Apple, 93
AT&T, xvii, 35, 36, 41, 62, 63, 113

B

Baldrige award, 70
Believing and doubting games, 138-139, 140
Bell Labs, 93
Bemkowski, Karen, 57
Benchmarking, 37, 41, 122, 123
Benchmarking, absolute, 41, 122, 123
Benchmarking, internal, 39
Betterment timeline principle, 16, 24
Boeing, 35
Bonaparte, Napoleon, 56
Brazil, 6, 45
Breakthrough thinking, viii, xi, xii, xvi, 5, 6, 8, 12, 15, 18, 20, 47, 61, 67, 72, 85, 86, 91, 97,100, 105, 106, 107, 110, 124, 126, 144

C

Canon, xvii
Champy, 65
Change, 51-56
Change, function of, 55
Change, managing, 51-52
Change, methodologies, 51
Change, models, 52, 54, 55, 62-63, 66

Chinese box, 14
Chuang-tse, 160
City Slickers, 51
Collison, William, 138
Commitment, 54, 104, 105, 106, 114, 140
Competitors, 30, 42
Concept creation, 27, 56, 67, 83, 91, 92, 93, 95, 97, 100, 101, 104, 111, 112, 124, 138, 146, 148, 149, 150
Concept engineering, 67, 83, 93, 94, 95, 97, 101, 104, 144, 146, 147, 149, 150
Concept focusing, 67, 83, 92, 93, 95, 97, 101, 104, 124, 146, 147, 148
Concept in, 67, 84, 91, 94, 95, 97, 101, 104, 116, 146, 147, 150
Concept management defined, 6
Concept management, stages of, 83-84
Concept management, xi, 36, 56, 61, 67, 82, 84, 85, 87, 95, 97, 101, 104, 107, 116, 121, 145, 146, 147, 150
Conflict reduction, 136, 137, 140
Conventional thinking, 12-14, 18, 20, 25, 26, 36, 89, 115, 121, 124, 145, 147, 150, 162
Core competencies, 107, 108, 114, 146
Core values, 107, 108, 114
Covey, Stephen R., 35, 43, 137
Creative Solution Finding, 12, 15
Critical resource, 120
Crosby, 73, 122
Cross-functional team, 94
Customer, 30, 36
Customer delight, 121
Customer satisfaction, 36, 121
Customer value-added, 36, 41
Cycle time management, 74

D

Deming, 4, 18, 62, 70, 73, 122
Descartes, 12, 22, 102, 145
Disney, Walt, 29
Disraeli, Benjamin, 64
Doyle Wilson Homebuilder, Inc., 4
Dreyfuss, Richard, 159

E

Efficiency, 40, 52
Einstein, Albert, 12
Electronic Data Interchange (EDI), 46
Empowerment, 32-33, 59, 105, 106, 112
Enterprise Resources Planning (ERP), 56, 73
Epictetus, 143

F

Fear, 53, 64, 143
FedEx, xvii, 30, 31, 33, 35, 43, 45, 86
Feedback, 60, 61
Financial improvements, 52
Financial performance measures, 37-38
Finite Capacity Scheduling (FCS), 73
Florida Power & Light, 62, 113
Flow charting (see Process mapping)
Ford, Henry, 30, 91
Ford Motor Company, 87, 95, 96
Fosdick, Harry Emerson, 73
Fuji Camera, 87
Future, 30, 42, 90, 95, 124, 150, 162

G

Gainsharing, 32-33, 59, 94, 105, 106, 112, 154
Globalization, 44
Globlocal, 135
GM, 35, 37
Goal, 105, 107, 108, 114, 120
God thinking, 12
Goethe, 33, 35
Grouping, 32
Guaranteed Fair Treatment Procedure (GFTP), 33

H

Hammer, 65
Hilton, 115
Hinkley, Gordon B., 141
Hitachi Wire Company, 26

Holland, Jeffrey R., 132
Holmes, Sherlock, 160, 162
Holonic view, 14, 19, 148
Hotel Nagoya Castle, 114, 115

I

IBM, 35, 38, 90, 93, 126
Ibuka, Masaru, 83, 93
In-Line Quality Control (ILQC), 61
Integration, 30, 44, 45, 102
Intel, 93, 138
Inventory, 40, 123
ISO 9000, xvii, 55, 69

J

John F. Kennedy High School, 159
Juran, 73, 122
Just-in-time (JIT), 55, 71, 91, 95, 96, 97, 125, 144, 145

K

KAIZEN, 55, 71
Kanban, 95, 96, 97, 145
KAO, 46, 100, 101
Keller, Helen, 3
Kimball, Spencer W., 139, 157
Kissinger, Henry, 153

L

Lincoln, Abraham, 48, 136
List-build-select (LBS) steps, 25

M

Machine view, 13-14
Malaysia, 30, 133, 138
Manufacturing Resources Planning (MRPII), 56, 73
Maxwell, Neal A., 159
Measurement, 31, 32, 44, 59, 61, 110, 123, 125, 126, 127, 156
Measurement/motivation relationship, 32, 36, 105, 106, 119-121, 126, 146
Micronesian, 157
Microsoft, 93, 94
MIS, 60
Mission, 44
Mitsubishi, xvii, 110, 111
Motivation (see measurement)

Index

Motorola, xvii
Mr. Holland's Opus, 159, 160
Multi-thinker, 13

N

Nadler, Gerald, x, xii, xv, 12, 113
Nagoya Institute of Technology, 96
National housing quality award, 4
NCR, 90, 93
Nippon Electric Corporation (NEC), 148-149

O

Ohmo, Taiichi, 96
Operational performance measures, 37-39
OR/MS Today, 65
Ownership, 32, 54, 62, 140, 160

P

Packer, Boyd K., 43
Panasonic, 110
Paradigm, 16, 27, 87, 162
Parieto principle, 60
Pauley, Linus, 133
People, 30, 44
People design principle, 16, 23
People-service-profit (P-S-P), 31
Performance, 30, 37
Plan of operation, 44
Plan-do-check-act, 62
Process mapping, 71
Process Reengineering (PR), xvii, 8, 55, 64, 65, 66, 72, 103
Productivity, 37, 39, 52, 120, 122, 126, 135
Pull thinking5, 7, 13, 19, 87, 90, 91, 144
Purpose, 14-16, 25, 130, 137, 156
Purpose expansion, 26, 92, 105, 110, 111, 112, 114, 115, 116, 136, 137, 146, 148, 149
Purpose hierarchy, 16, 18, 19
Purpose-Target-Results Approach (PTR Approach), 25-26, 27
Purposes principle, 16, 17-19
Push thinking, 5, 7, 13, 19, 87, 89, 90, 91

Q

QS 9000, 126
Quality, 37, 39, 52, 121
Quality circles, 31, 55, 72, 120, 132, 145
Quality council, 58-59, 60, 105, 108, 109, 110, 114, 132, 133, 140

Quality Functional Deployment (QFD), 52, 55, 69
Quiz Show, 33

R

Regularity concept, 100, 101
Result, 25
Road map, 5, 7
Root cause, 5, 85, 87, 89, 92, 136, 145, 147, 148
Ruskin, John, 133

S

Saturn, 37
Schumacher, E. F., 117
Seki, 26
Semco, 6, 34, 45, 46, 93, 96
Semler, Ricardo, 6, 34, 46, 93, 96, 106
Shaw, George Bernard, 67
Shikumi Zukuri, 144
Shingo award, 70
Showa Ceramic Company, 15, 16, 138
Six sigma region, 100, 101
Sliders, 83
Small is Beautiful, 113, 117
SME/CASA, 65
Smith, Frederick W., 30
Solution-after-next principle (SAN), 16, 19, 20, 24, 25, 27
Sony Corporation, 83
Statistical process control (SPC), 31, 56, 61, 72, 89, 120
Strategy, 44
System, 14
System view, 13-14, 27
Systematic Problem Solving (SPS), 61, 62-63, 94, 105, 110, 112, 113, 114
Systems principle, 16, 20

T

3M Company, 86
T-Model, 62, 63, 64
Target, 25
Target objectives, 5
Teaming (see Teams)
Teams, 32, 105, 106, 108, 109, 110, 112, 114, 116, 132, 140, 156
Teams, functions of, 60
Teams, three "P" teams, 59
The Goal, 113
The Mission Statement Book, 108

The Seven Habits of Highly Effective People, 113
The Te of Piglet, 160, 162
Throughput, 40
Time-to-market (TTM), 74
Tokyo Hotel, 115
Toshiba, 86
Total Quality Control (TQC), 56, 61, 89, 120
Toyoda, Kiichiro, 96
Toyoda, Sakichi, 96
Toyota Central Research Labs, 143, 144, 147
Toyota Motors, 96, 143, 145
Toyota Power Company, 26
Toyota production system, 145
Toyota Weaving Company, 96
Toyota, xvii, 4, 71, 91, 123, 149
TQM (Total Quality Management), xvii, 8, 33, 52, 55, 56, 61, 65, 66, 67, 72, 73, 74, 86, 93, 94, 97, 105, 106, 112, 114, 150
TQM, operational (procedural) elements of, 56, 58
TQM, philosophical elements of, 56, 57
TQM, project implementation steps, 59-60
TQM, shortcomings of, 61
TQM, summary, 64
Training, 105, 107, 111, 114, 126
Trenton Forging Company, 126
TRIZ, 5, 89
Tuttle, Tom, 125

U

Uniqueness principle, 16
United States Dept. of Defense, 101
UPS, 126

V

Value-adding, 40, 103, 122, 134, 135
Van Gogh, Vincent, xi
Vision, 44

W

Wal-Mart, 46
Walt Disney Corporation, 66, 159, 160
Waste, 40, 111, 153, 156
Wholeness, 14-16, 27, 137
Wilson, Doyle, 4
Windows, 94
World class, 36, 47, 100, 116, 127, 132, 134, 140
World class management, xi, xii, xvi, 5, 6, 8, 30, 41, 47, 67, 86, 91, 94, 95, 97, 106, 107, 114, 124, 150
World Class Manager, 27, 45, 86 92, 112, 113, 132, 134, 160

X

Xerox, xvii